Fundamentals of Machine Learning

Fundamentals of Machine Learning

Edited by
Percy Vaughn

Larsen & Keller
www.larsen-keller.com

Fundamentals of Machine Learning
Edited by Percy Vaughn
ISBN: 978-1-63549-679-6 (Hardback)

 Larsen & Keller

Published by Larsen and Keller Education,
5 Penn Plaza,
19th Floor,
New York, NY 10001, USA

Cataloging-in-Publication Data

Fundamentals of machine learning / edited by Percy Vaughn.
    p. cm.
Includes bibliographical references and index.
ISBN 978-1-63549-679-6
1. Machine learning. 2. Artificial intelligence. 3. Machine theory. I. Vaughn, Percy.
Q325.5 .F86 2018
006.31--dc23

For more information regarding Larsen and Keller Education and its products, please visit the publisher's website www.larsen-keller.com

Table of Contents

Preface

Machine learning refers to the ability of computers to recognize patterns and arrive at decisions based on such assessment. It helps in a better analysis of data sets and thereby determining better predictions. The different concepts used under this field are deep learning, decision tree learning, clustering, reinforcement learning, sparse dictionary learning, rule-based machine learning, etc. This book is a valuable compilation of topics, ranging from the basic to the most complex theories and principles in the field of machine learning. The topics included in it are of utmost significance and bound to provide incredible insights to readers. The textbook is appropriate for those seeking detailed information in this area.

To facilitate a deeper understanding of the contents of this book a short introduction of every chapter is written below:

Chapter 1- Machine learning has become a part of our everyday life. It studies algorithms which help in studying and making predictions on data. Machine learning is employed to perform difficult tasks like email filtering, detecting network intruders, learning to rank and computer vision. The chapter on machine learning offers an insightful focus, keeping in mind the complex subject matter.

Chapter 2- Supervised learning infers tasks from labeled data while unsupervised learning uses unlabeled data for computing a function. Semi-supervised learning, structured prediction, bias–variance tradeoff, competitive learning and autoencoder are some important topics related to the subject matter. The major components of supervised and unsupervised learning are discussed in this chapter.

Chapter 3- The methods of machine learning are logic learning machine, online machine learning, rule-based machine learning, multiple kernel learning and temporal difference learning. Logic learning machine is a method which has its basis on intelligible rules. This chapter discusses the methods of machine learning in a critical manner providing key analysis to the subject matter.

Chapter 4- Important machine learning algorithms include expectation-maximization algorithm, structured kNN, wake-sleep algorithm, etc. Structured kNN algorithms generalizes the kNN classifier. The topics discussed in the chapter are of great importance to broaden the existing knowledge on machine learning.

Chapter 5- Meta leaning is a branch of machine learning which applies automatic learning algorithms to meta-data. This process is used to understand how automatic leaning can be made more flexible and can be used in solving difficult learning problems. This chapter is a compilation of the various allied fields of machine learning that form an integral part of the broader subject matter.

Chapter 6- Artificial neural networks are computing systems that are motivated by biological neural networks. The system is based on units called artificial neurons. Information is sent from one neuron to another and the receiving neuron processes it. This chapter will provide an integrated understanding of artificial neural networks.

I would like to share the credit of this book with my editorial team who worked tirelessly on this book. I owe the completion of this book to the never-ending support of my family, who supported me throughout the project.

Editor

An Introduction to Machine Learning

Machine learning has become a part of our everyday life. It studies algorithms which help in studying and making predictions on data. Machine learning is employed to perform difficult tasks like email filtering, detecting network intruders, learning to rank and computer vision. The chapter on machine learning offers an insightful focus, keeping in mind the complex subject matter.

Machine Learning

Machine learning is the subfield of computer science that, according to Arthur Samuel in 1959, gives "computers the ability to learn without being explicitly programmed." Evolved from the study of pattern recognition and computational learning theory in artificial intelligence, machine learning explores the study and construction of algorithms that can learn from and make predictions on data – such algorithms overcome following strictly static program instructions by making data-driven predictions or decisions, through building a model from sample inputs. Machine learning is employed in a range of computing tasks where designing and programming explicit algorithms with good performance is difficult or infeasible; example applications include email filtering, detection of network intruders or malicious insiders working towards a data breach, optical character recognition (OCR), learning to rank, and computer vision.

Machine learning is closely related to (and often overlaps with) computational statistics, which also focuses on prediction-making through the use of computers. It has strong ties to mathematical optimization, which delivers methods, theory and application domains to the field. Machine learning is sometimes conflated with data mining, where the latter subfield focuses more on exploratory data analysis and is known as unsupervised learning. Machine learning can also be unsupervised and be used to learn and establish baseline behavioral profiles for various entities and then used to find meaningful anomalies.

Within the field of data analytics, machine learning is a method used to devise complex models and algorithms that lend themselves to prediction; in commercial use, this is known as predictive analytics. These analytical models allow researchers, data scientists, engineers, and analysts to "produce reliable, repeatable decisions and results" and uncover "hidden insights" through learning from historical relationships and trends in the data.

As of 2016, machine learning is a buzzword, and according to the Gartner hype cycle of 2016, at its peak of inflated expectations. Because finding patterns is hard, often not enough training data is available, and also because of the high expectations it often fails to deliver.

Overview

Tom M. Mitchell provided a widely quoted, more formal definition: "A computer program is said to learn from experience E with respect to some class of tasks T and performance measure P if its performance at tasks in T, as measured by P, improves with experience E." This definition is notable for its defining machine learning in fundamentally operational rather than cognitive terms, thus following Alan Turing's proposal in his paper "Computing Machinery and Intelligence", that the question "Can machines think?" be replaced with the question "Can machines do what we (as thinking entities) can do?". In the proposal he explores the various characteristics that could be possessed by a *thinking machine* and the various implications in constructing one.

Types of Problems and Tasks

Machine learning tasks are typically classified into three broad categories, depending on the nature of the learning "signal" or "feedback" available to a learning system. These are

- Supervised learning: The computer is presented with example inputs and their desired outputs, given by a "teacher", and the goal is to learn a general rule that maps inputs to outputs.

- Unsupervised learning: No labels are given to the learning algorithm, leaving it on its own to find structure in its input. Unsupervised learning can be a goal in itself (discovering hidden patterns in data) or a means towards an end (feature learning).

- Reinforcement learning: A computer program interacts with a dynamic environment in which it must perform a certain goal (such as driving a vehicle or playing a game against an opponent). The program is provided feedback in terms of rewards and punishments as it navigates its problem space.

Between supervised and unsupervised learning is semi-supervised learning, where the teacher gives an incomplete training signal: a training set with some (often many) of the target outputs missing. Transduction is a special case of this principle where the entire set of problem instances is known at learning time, except that part of the targets are missing.

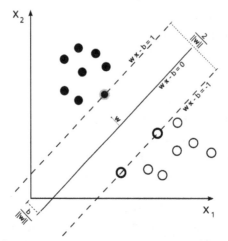

A support vector machine is a classifier that divides its input space into two regions, separated by a linear boundary. Here, it has learned to distinguish black and white circles.

Among other categories of machine learning problems, learning to learn learns its own inductive bias based on previous experience. Developmental learning, elaborated for robot learning, generates its own sequences (also called curriculum) of learning situations to cumulatively acquire repertoires of novel skills through autonomous self-exploration and social interaction with human teachers and using guidance mechanisms such as active learning, maturation, motor synergies, and imitation.

Another categorization of machine learning tasks arises when one considers the desired *output* of a machine-learned system:

- In classification, inputs are divided into two or more classes, and the learner must produce a model that assigns unseen inputs to one or more (multi-label classification) of these classes. This is typically tackled in a supervised way. Spam filtering is an example of classification, where the inputs are email (or other) messages and the classes are "spam" and "not spam".

- In regression, also a supervised problem, the outputs are continuous rather than discrete.

- In clustering, a set of inputs is to be divided into groups. Unlike in classification, the groups are not known beforehand, making this typically an unsupervised task.

- Density estimation finds the distribution of inputs in some space.

- Dimensionality reduction simplifies inputs by mapping them into a lower-dimensional space. Topic modeling is a related problem, where a program is given a list of human language documents and is tasked to find out which documents cover similar topics.

History and Relationships to Other Fields

As a scientific endeavour, machine learning grew out of the quest for artificial intelligence. Already in the early days of AI as an academic discipline, some researchers were interested in having machines learn from data. They attempted to approach the problem with various symbolic methods, as well as what were then termed "neural networks"; these were mostly perceptrons and other models that were later found to be reinventions of the generalized linear models of statistics. Probabilistic reasoning was also employed, especially in automated medical diagnosis.

However, an increasing emphasis on the logical, knowledge-based approach caused a rift between AI and machine learning. Probabilistic systems were plagued by theoretical and practical problems of data acquisition and representation. By 1980, expert systems had come to dominate AI, and statistics was out of favor. Work on symbolic/knowledge-based learning did continue within AI, leading to inductive logic programming, but the more statistical line of research was now outside the field of AI proper, in pattern recognition and information retrieval. Neural networks research had been abandoned by AI and computer science around the same time. This line, too, was continued outside the AI/CS field, as "connectionism", by researchers from other disciplines including Hopfield, Rumelhart and Hinton. Their main success came in the mid-1980s with the reinvention of backpropagation.

Machine learning, reorganized as a separate field, started to flourish in the 1990s. The field changed its goal from achieving artificial intelligence to tackling solvable problems of a practical nature. It shifted focus away from the symbolic approaches it had inherited from AI, and toward methods and models borrowed from statistics and probability theory. It also benefited from the increasing availability of digitized information, and the possibility to distribute that via the Internet.

Machine learning and data mining often employ the same methods and overlap significantly, but while machine learning focuses on prediction, based on *known* properties learned from the training data, data mining focuses on the discovery of (previously) *unknown* properties in the data (this is the analysis step of Knowledge Discovery in Databases). Data mining uses many machine learning methods, but with different goals; on the other hand, machine learning also employs data mining methods as "unsupervised learning" or as a preprocessing step to improve learner accuracy. Much of the confusion between these two research communities (which do often have separate conferences and separate journals, ECML PKDD being a major exception) comes from the basic assumptions they work with: in machine learning, performance is usually evaluated with respect to the ability to *reproduce known* knowledge, while in Knowledge Discovery and Data Mining (KDD) the key task is the discovery of previously *unknown* knowledge. Evaluated with respect to known knowledge, an uninformed (unsupervised) method will easily be outperformed by other supervised methods, while in a typical KDD task, supervised methods cannot be used due to the unavailability of training data.

Machine learning also has intimate ties to optimization: many learning problems are formulated as minimization of some loss function on a training set of examples. Loss functions express the discrepancy between the predictions of the model being trained and the actual problem instances (for example, in classification, one wants to assign a label to instances, and models are trained to correctly predict the pre-assigned labels of a set of examples). The difference between the two fields arises from the goal of generalization: while optimization algorithms can minimize the loss on a training set, machine learning is concerned with minimizing the loss on unseen samples.

Relation to Statistics

Machine learning and statistics are closely related fields. According to Michael I. Jordan, the ideas of machine learning, from methodological principles to theoretical tools, have had a long pre-history in statistics. He also suggested the term data science as a placeholder to call the overall field.

Leo Breiman distinguished two statistical modelling paradigms: data model and algorithmic model, wherein 'algorithmic model' means more or less the machine learning algorithms like Random forest.

Some statisticians have adopted methods from machine learning, leading to a combined field that they call *statistical learning*.

Theory

A core objective of a learner is to generalize from its experience. Generalization in this context is the ability of a learning machine to perform accurately on new, unseen examples/tasks after having experienced a learning data set. The training examples come from some generally unknown probability distribution (considered representative of the space of occurrences) and the learner has to build a general model about this space that enables it to produce sufficiently accurate predictions in new cases.

The computational analysis of machine learning algorithms and their performance is a branch of theoretical computer science known as computational learning theory. Because training sets are finite and the future is uncertain, learning theory usually does not yield guarantees of the performance of algorithms. Instead, probabilistic bounds on the performance are quite common. The bias–variance decomposition is one way to quantify generalization error.

For the best performance in the context of generalization, the complexity of the hypothesis should match the complexity of the function underlying the data. If the hypothesis is less complex than the function, then the model has underfit the data. If the complexity of the model is increased in response, then the training error decreases. But if the hypothesis is too complex, then the model is subject to overfitting and generalization will be poorer.

In addition to performance bounds, computational learning theorists study the time complexity and feasibility of learning. In computational learning theory, a computation is considered feasible if it can be done in polynomial time. There are two kinds of time complexity results. Positive results show that a certain class of functions can be learned in polynomial time. Negative results show that certain classes cannot be learned in polynomial time.

Decision Tree Learning

Decision tree learning uses a decision tree as a predictive model, which maps observations about an item to conclusions about the item's target value.

Association Rule Learning

Association rule learning is a method for discovering interesting relations between variables in large databases.

Artificial Neural Networks

An artificial neural network (ANN) learning algorithm, usually called "neural network" (NN), is a learning algorithm that is inspired by the structure and functional aspects of biological neural networks. Computations are structured in terms of an interconnected group of artificial neurons, processing information using a connectionist approach to computation. Modern neural networks are non-linear statistical data modeling tools. They are usually used to model complex relationships between inputs and outputs, to find patterns in data, or to capture the statistical structure in an unknown joint probability distribution between observed variables.

Deep Learning

Falling hardware prices and the development of GPUs for personal use in the last few years have contributed to the development of the concept of deep learning which consists of multiple hidden layers in an artificial neural network. This approach tries to model the way the human brain processes light and sound into vision and hearing. Some successful applications of deep learning are computer vision and speech recognition.

Inductive Logic Programming

Inductive logic programming (ILP) is an approach to rule learning using logic programming as a uniform representation for input examples, background knowledge, and hypotheses. Given an encoding of the known background knowledge and a set of examples represented as a logical database of facts, an ILP system will derive a hypothesized logic program that entails all positive and no negative examples. Inductive programming is a related field that considers any kind of programming languages for representing hypotheses (and not only logic programming), such as functional programs.

Support Vector Machines

Support vector machines (SVMs) are a set of related supervised learning methods used for classification and regression. Given a set of training examples, each marked as belonging to one of two categories, an SVM training algorithm builds a model that predicts whether a new example falls into one category or the other.

Clustering

Cluster analysis is the assignment of a set of observations into subsets (called *clusters*) so that observations within the same cluster are similar according to some predesignated criterion or criteria, while observations drawn from different clusters are dissimilar. Different clustering techniques make different assumptions on the structure of the data, often defined by some *similarity metric* and evaluated for example by *internal compactness* (similarity between members of the same cluster) and *separation* between different clusters. Other methods are based on *estimated density* and *graph connectivity*. Clustering is a method of unsupervised learning, and a common technique for statistical data analysis.

Bayesian Networks

A Bayesian network, belief network or directed acyclic graphical model is a probabilistic graphical model that represents a set of random variables and their conditional independencies via a directed acyclic graph (DAG). For example, a Bayesian network could represent the probabilistic relationships between diseases and symptoms. Given symptoms, the network can be used to compute the probabilities of the presence of various diseases. Efficient algorithms exist that perform inference and learning.

Reinforcement Learning

Reinforcement learning is concerned with how an *agent* ought to take *actions* in an *environment* so as to maximize some notion of long-term *reward*. Reinforcement learning algorithms attempt to find a *policy* that maps *states* of the world to the actions the agent ought to take in those states. Reinforcement learning differs from the supervised learning problem in that correct input/output pairs are never presented, nor sub-optimal actions explicitly corrected.

Representation Learning

Several learning algorithms, mostly unsupervised learning algorithms, aim at discovering better representations of the inputs provided during training. Classical examples include principal components analysis and cluster analysis. Representation learning algorithms often attempt to preserve the information in their input but transform it in a way that makes it useful, often as a pre-processing step before performing classification or predictions, allowing reconstruction of the inputs coming from the unknown data generating distribution, while not being necessarily faithful for configurations that are implausible under that distribution.

Manifold learning algorithms attempt to do so under the constraint that the learned representation is low-dimensional. Sparse coding algorithms attempt to do so under the constraint that the learned representation is sparse (has many zeros). Multilinear subspace learning algorithms aim to learn low-dimensional representations directly from tensor representations for multidimensional data, without reshaping them into (high-dimensional) vectors. Deep learning algorithms discover multiple levels of representation, or a hierarchy of features, with higher-level, more abstract features defined in terms of (or generating) lower-level features. It has been argued that an intelligent machine is one that learns a representation that disentangles the underlying factors of variation that explain the observed data.

Similarity and Metric Learning

In this problem, the learning machine is given pairs of examples that are considered similar and pairs of less similar objects. It then needs to learn a similarity function (or a distance metric function) that can predict if new objects are similar. It is sometimes used in Recommendation systems.

Sparse Dictionary Learning

In this method, a datum is represented as a linear combination of basis functions, and the coefficients are assumed to be sparse. Let x be a d-dimensional datum, D be a d by n matrix, where each column of D represents a basis function. r is the coefficient to represent x using D. Mathematically, sparse dictionary learning means solving $x \approx Dr$ where r is sparse. Generally speaking, n is assumed to be larger than d to allow the freedom for a sparse representation.

Learning a dictionary along with sparse representations is strongly NP-hard and also difficult to solve approximately. A popular heuristic method for sparse dictionary learning is K-SVD.

Sparse dictionary learning has been applied in several contexts. In classification, the problem is to determine which classes a previously unseen datum belongs to. Suppose a dictionary for each class has already been built. Then a new datum is associated with the class such that it's best sparsely represented by the corresponding dictionary. Sparse dictionary learning has also been applied in image de-noising. The key idea is that a clean image patch can be sparsely represented by an image dictionary, but the noise cannot.

Genetic Algorithms

A genetic algorithm (GA) is a search heuristic that mimics the process of natural selection, and uses methods such as mutation and crossover to generate new genotype in the hope of finding good solutions to a given problem. In machine learning, genetic algorithms found some uses in the 1980s and 1990s. Vice versa, machine learning techniques have been used to improve the performance of genetic and evolutionary algorithms.

Rule-based Machine Learning

Rule-based machine learning is a general term for any machine learning method that identifies, learns, or evolves `rules' to store, manipulate or apply, knowledge. The defining characteristic of a rule-based machine learner is the identification and utilization of a set of relational rules that collectively represent the knowledge captured by the system. This is in contrast to other machine learners that commonly identify a singular model that can be universally applied to any instance in order to make a prediction. Rule-based machine learning approaches include learning classifier systems, association rule learning, and artificial immune systems.

Learning Classifier Systems

Learning classifier systems (LCS) are a family of rule-based machine learning algorithms that combine a discovery component (e.g. typically a genetic algorithm) with a learning component (performing either supervised learning, reinforcement learning, or unsupervised learning). They seek to identify a set of context-dependent rules that collectively store and apply knowledge in a piecewise manner in order to make predictions.

Applications

Applications for machine learning include:

- Adaptive websites
- Affective computing
- Bioinformatics
- Brain-machine interfaces
- Cheminformatics
- Classifying DNA sequences
- Computational anatomy
- Computer vision, including object recognition
- Detecting credit card fraud
- Game playing
- Information retrieval
- Internet fraud detection
- Marketing
- Machine learning control
- Machine perception
- Medical diagnosis

- Economics

- Natural language processing

- Natural language understanding

- Optimization and metaheuristic

- Online advertising

- Recommender systems

- Robot locomotion

- Search engines

- Sentiment analysis (or opinion mining)

- Sequence mining

- Software engineering

- Speech and handwriting recognition

- Financial market analysis

- Structural health monitoring

- Syntactic pattern recognition

- User behavior analytics

- Translation

In 2006, the online movie company Netflix held the first "Netflix Prize" competition to find a program to better predict user preferences and improve the accuracy on its existing Cinematch movie recommendation algorithm by at least 10%. A joint team made up of researchers from AT&T Labs-Research in collaboration with the teams Big Chaos and Pragmatic Theory built an ensemble model to win the Grand Prize in 2009 for $1 million. Shortly after the prize was awarded, Netflix realized that viewers' ratings were not the best indicators of their viewing patterns ("everything is a recommendation") and they changed their recommendation engine accordingly.

In 2012, co-founder of Sun Microsystems Vinod Khosla predicted that 80% of medical doctors jobs would be lost in the next two decades to automated machine learning medical diagnostic software.

In 2014, it has been reported that a machine learning algorithm has been applied in Art History to study fine art paintings, and that it may have revealed previously unrecognized influences between artists.

Model Assessments

Classification machine learning models can be validated by accuracy estimation techniques like the Holdout method, which splits the data in a training and test set (conventionally 2/3 training set and 1/3 test set designation) and evaluates the performance of the training model on the test set. In comparison, the N-fold-cross-validation method randomly splits the data in k subsets where the k-1 instances of the data are used to train the model while the kth instance is used to test the predictive ability of the training model. In addition to the holdout and cross-validation methods, bootstrap, which samples n instances with replacement from the dataset, can be used to assess model accuracy.

In addition to overall accuracy, investigators frequently report sensitivity and specificity (True Positive Rate: TPR and True Negative Rate: TNR, respectively) meaning True Positive Rate (TPR) and True Negative Rate (TNR) respectively. Similarly, investigators sometimes report the False Positive Rate (FPR) as well as the False Negative Rate (FNR). However, these rates are ratios that fail to reveal their numerators and denominators. The Total Operating Characteristic (TOC) is an effective method to express a model's diagnostic ability. TOC shows the numerators and denominators of the previously mentioned rates, thus TOC provides more information than the commonly used Receiver operating characteristic (ROC) and ROC's associated Area Under the Curve (AUC).

Ethics

Machine Learning poses a host of ethical questions. Systems which are trained on datasets collected with biases may exhibit these biases upon use, thus digitizing cultural prejudices. Responsible collection of data thus is a critical part of machine learning.

Because language contains biases, machines trained on language corpora will necessarily also learn bias.

Software

Software suites containing a variety of machine learning algorithms include the following :

Free and Open-source Software

- CNTK
- Deeplearning4j
- dlib
- ELKI
- GNU Octave
- H2O
- Mahout
- Mallet
- mlpy
- MLPACK
- MOA (Massive Online Analysis)
- MXNet
- ND4J: ND arrays for Java
- NuPIC
- OpenAI Gym
- OpenAI Universe
- OpenNN
- Orange

- R
- scikit-learn
- Shogun
- SMILE
- SparkML

- TensorFlow
- Torch
- Yooreeka
- Weka

Proprietary Software with Free and Open-source Editions

- KNIME
- RapidMiner

Proprietary Software

- Amazon Machine Learning
- Angoss KnowledgeSTUDIO
- Ayasdi
- Google Prediction API
- IBM SPSS Modeler
- KXEN Modeler
- LIONsolver
- Mathematica
- MATLAB
- Microsoft Azure Machine Learning

- Neural Designer
- NeuroSolutions
- Oracle Data Mining
- RCASE
- SAS Enterprise Miner
- SequenceL
- Skymind
- Splunk
- STATISTICA Data Miner

Pattern Recognition

Pattern recognition is a branch of machine learning that focuses on the recognition of patterns and regularities in data, although it is in some cases considered to be nearly synonymous with machine learning. Pattern recognition systems are in many cases trained from labeled "training" data (supervised learning), but when no labeled data are available other algorithms can be used to discover previously unknown patterns (unsupervised learning).

The terms pattern recognition, machine learning, data mining and knowledge discovery in databases (KDD) are hard to separate, as they largely overlap in their scope. Machine learning is the common term for supervised learning methods and originates from artificial intelligence, whereas

KDD and data mining have a larger focus on unsupervised methods and stronger connection to business use. Pattern recognition has its origins in engineering, and the term is popular in the context of computer vision: a leading computer vision conference is named Conference on Computer Vision and Pattern Recognition. In pattern recognition, there may be a higher interest to formalize, explain and visualize the pattern, while machine learning traditionally focuses on maximizing the recognition rates. Yet, all of these domains have evolved substantially from their roots in artificial intelligence, engineering and statistics, and they've become increasingly similar by integrating developments and ideas from each other.

In machine learning, pattern recognition is the assignment of a label to a given input value. In statistics, discriminant analysis was introduced for this same purpose in 1936. An example of pattern recognition is classification, which attempts to assign each input value to one of a given set of *classes* (for example, determine whether a given email is "spam" or "non-spam"). However, pattern recognition is a more general problem that encompasses other types of output as well. Other examples are regression, which assigns a real-valued output to each input; sequence labeling, which assigns a class to each member of a sequence of values (for example, part of speech tagging, which assigns a part of speech to each word in an input sentence); and parsing, which assigns a parse tree to an input sentence, describing the syntactic structure of the sentence.

Pattern recognition algorithms generally aim to provide a reasonable answer for all possible inputs and to perform "most likely" matching of the inputs, taking into account their statistical variation. This is opposed to *pattern matching* algorithms, which look for exact matches in the input with pre-existing patterns. A common example of a pattern-matching algorithm is regular expression matching, which looks for patterns of a given sort in textual data and is included in the search capabilities of many text editors and word processors. In contrast to pattern recognition, pattern matching is generally not considered a type of machine learning, although pattern-matching algorithms (especially with fairly general, carefully tailored patterns) can sometimes succeed in providing similar-quality output of the sort provided by pattern-recognition algorithms.

Overview

Pattern recognition is generally categorized according to the type of learning procedure used to generate the output value. *Supervised learning* assumes that a set of *training data* (the *training set*) has been provided, consisting of a set of instances that have been properly labeled by hand with the correct output. A learning procedure then generates a *model* that attempts to meet two sometimes conflicting objectives: Perform as well as possible on the training data, and generalize as well as possible to new data (usually, this means being as simple as possible, for some technical definition of "simple", in accordance with Occam's Razor, discussed below). Unsupervised learning, on the other hand, assumes training data that has not been hand-labeled, and attempts to find inherent patterns in the data that can then be used to determine the correct output value for new data instances. A combination of the two that has recently been explored is semi-supervised learning, which uses a combination of labeled and unlabeled data (typically a small set of labeled data combined with a large amount of unlabeled data). Note that in cases of unsupervised learning, there may be no training data at all to speak of; in other words, the data to be labeled *is* the training data.

Note that sometimes different terms are used to describe the corresponding supervised and unsupervised learning procedures for the same type of output. For example, the unsupervised equivalent of classification is normally known as *clustering*, based on the common perception of the task as involving no training data to speak of, and of grouping the input data into *clusters* based on some inherent similarity measure (e.g. the distance between instances, considered as vectors in a multi-dimensional vector space), rather than assigning each input instance into one of a set of pre-defined classes. Note also that in some fields, the terminology is different: For example, in community ecology, the term "classification" is used to refer to what is commonly known as "clustering".

The piece of input data for which an output value is generated is formally termed an *instance*. The instance is formally described by a vector of *features*, which together constitute a description of all known characteristics of the instance. (These feature vectors can be seen as defining points in an appropriate multidimensional space, and methods for manipulating vectors in vector spaces can be correspondingly applied to them, such as computing the dot product or the angle between two vectors.) Typically, features are either categorical (also known as nominal, i.e., consisting of one of a set of unordered items, such as a gender of "male" or "female", or a blood type of "A", "B", "AB" or "O"), ordinal (consisting of one of a set of ordered items, e.g., "large", "medium" or "small"), integer-valued (e.g., a count of the number of occurrences of a particular word in an email) or real-valued (e.g., a measurement of blood pressure). Often, categorical and ordinal data are grouped together; likewise for integer-valued and real-valued data. Furthermore, many algorithms work only in terms of categorical data and require that real-valued or integer-valued data be *discretized* into groups (e.g., less than 5, between 5 and 10, or greater than 10).

Probabilistic Classifiers

Many common pattern recognition algorithms are *probabilistic* in nature, in that they use statistical inference to find the best label for a given instance. Unlike other algorithms, which simply output a "best" label, often probabilistic algorithms also output a probability of the instance being described by the given label. In addition, many probabilistic algorithms output a list of the N-best labels with associated probabilities, for some value of N, instead of simply a single best label. When the number of possible labels is fairly small (e.g., in the case of classification), N may be set so that the probability of all possible labels is output. Probabilistic algorithms have many advantages over non-probabilistic algorithms:

- They output a confidence value associated with their choice. (Note that some other algorithms may also output confidence values, but in general, only for probabilistic algorithms is this value mathematically grounded in probability theory. Non-probabilistic confidence values can in general not be given any specific meaning, and only used to compare against other confidence values output by the same algorithm.)

- Correspondingly, they can *abstain* when the confidence of choosing any particular output is too low.

- Because of the probabilities output, probabilistic pattern-recognition algorithms can be more effectively incorporated into larger machine-learning tasks, in a way that partially or completely avoids the problem of *error propagation*.

Number of Important Feature Variables

Feature selection algorithms attempt to directly prune out redundant or irrelevant features. A general introduction to feature selection which summarizes approaches and challenges, has been given. The complexity of feature-selection is, because of its non-monotonous character, an optimization problem where given a total of n features the powerset consisting of all $2^n - 1$ subsets of features need to be explored. The Branch-and-Bound algorithm does reduce this complexity but is intractable for medium to large values of the number of available features n. For a large-scale comparison of feature-selection algorithms.

Techniques to transform the raw feature vectors (feature extraction) are sometimes used prior to application of the pattern-matching algorithm. For example, feature extraction algorithms attempt to reduce a large-dimensionality feature vector into a smaller-dimensionality vector that is easier to work with and encodes less redundancy, using mathematical techniques such as principal components analysis (PCA). The distinction between feature selection and feature extraction is that the resulting features after feature extraction has taken place are of a different sort than the original features and may not easily be interpretable, while the features left after feature selection are simply a subset of the original features.

Problem Statement (Supervised Version)

Formally, the problem of supervised pattern recognition can be stated as follows: Given an unknown function $g : \mathcal{X} \to \mathcal{Y}$ (the *ground truth*) that maps input instances $\mathbf{x} \in \mathcal{X}$ to output labels $y \in \mathcal{Y}$, along with training data $\mathbf{D} = \{(\mathbf{x}_1, y_1), \ldots, (\mathbf{x}_n, y_n)\}$ assumed to represent accurate examples of the mapping, produce a function $h : \mathcal{X} \to \mathcal{Y}$ that approximates as closely as possible the correct mapping g. (For example, if the problem is filtering spam, then $\mathbf{x}_i$ is some representation of an email and y is either "spam" or "non-spam"). In order for this to be a well-defined problem, "approximates as closely as possible" needs to be defined rigorously. In decision theory, this is defined by specifying a loss function or cost function that assigns a specific value to "loss" resulting from producing an incorrect label. The goal then is to minimize the expected loss, with the expectation taken over the probability distribution of $\mathcal{X}$. In practice, neither the distribution of $\mathcal{X}$ nor the ground truth function $g : \mathcal{X} \to \mathcal{Y}$ are known exactly, but can be computed only empirically by collecting a large number of samples of $\mathcal{X}$ and hand-labeling them using the correct value of $\mathcal{Y}$ (a time-consuming process, which is typically the limiting factor in the amount of data of this sort that can be collected). The particular loss function depends on the type of label being predicted. For example, in the case of classification, the simple zero-one loss function is often sufficient. This corresponds simply to assigning a loss of 1 to any incorrect labeling and implies that the optimal classifier minimizes the error rate on independent test data (i.e. counting up the fraction of instances that the learned function $h : \mathcal{X} \to \mathcal{Y}$ labels wrongly, which is equivalent to maximizing the number of correctly classified instances). The goal of the learning procedure is then to minimize the error rate (maximize the correctness) on a "typical" test set.

For a probabilistic pattern recognizer, the problem is instead to estimate the probability of each possible output label given a particular input instance, i.e., to estimate a function of the form

$$p(\text{label} \mid x, \theta) = f(x; \theta)$$

where the feature vector input is $\mathbf{x}$, and the function f is typically parameterized by some parameters θ. In a discriminative approach to the problem, f is estimated directly. In a generative approach, however, the inverse probability $p(\mathbf{x}\,|\,\text{label})$ is instead estimated and combined with the prior probability $p(\text{label}\,|\,\theta)$ using Bayes' rule, as follows:

$$p(\text{label}\,|\,\mathbf{x},\theta) = \frac{p(\mathbf{x}\,|\,\text{label},\theta)p(\text{label}|\theta)}{\sum_{L\in\text{all labels}}p(\mathbf{x}\,|\,L)p(L\,|\,\theta)}.$$

When the labels are continuously distributed (e.g., in regression analysis), the denominator involves integration rather than summation:

$$p(\text{label}\,|\,\mathbf{x},\theta) = \frac{p(\mathbf{x}\,|\,\text{label},\theta)p(\text{label}|\theta)}{\int_{L\in\text{all labels}}p(\mathbf{x}\,|\,L)p(L\,|\,\theta)\,d\,L}.$$

The value of θ is typically learned using maximum a posteriori (MAP) estimation. This finds the best value that simultaneously meets two conflicting objects: To perform as well as possible on the training data (smallest error-rate) and to find the simplest possible model. Essentially, this combines maximum likelihood estimation with a regularization procedure that favors simpler models over more complex models. In a Bayesian context, the regularization procedure can be viewed as placing a prior probability $p(\theta)$ on different values of θ. Mathematically:

$$\theta^* = \arg\max_{\theta} p(\theta\,|\,D)$$

where θ^* is the value used for θ in the subsequent evaluation procedure, and $p(\theta\,|\,D)$, the posterior probability of θ, is given by

$$p(\theta\,|\,D) = \left[\prod_{i=1}^{n}p(y_i\,|\,x_i,\theta)\right]p(\theta).$$

In the Bayesian approach to this problem, instead of choosing a single parameter vector θ^*, the probability of a given label for a new instance $\mathbf{x}$ is computed by integrating over all possible values of θ, weighted according to the posterior probability:

$$p(\text{label}\,|\,\mathbf{x}) = \int p(\text{label}\,|\,\mathbf{x},\theta)p(\theta\,|\,D)\,d\,\theta.$$

Frequentist or Bayesian Approach to Pattern Recognition

The first pattern classifier – the linear discriminant presented by Fisher – was developed in the Frequentist tradition. The frequentist approach entails that the model parameters are considered unknown, but objective. The parameters are then computed (estimated) from the collected data. For the linear discriminant, these parameters are precisely the mean vectors and the Covariance matrix. Also the probability of each class $p(\text{label}\,|\,\theta)$ is estimated from the collected dataset. Note that the usage of 'Bayes rule' in a pattern classifier does not make the classification approach Bayesian.

Bayesian statistics has its origin in Greek philosophy where a distinction was already made between the 'a priori' and the 'a posteriori' knowledge. Later Kant defined his distinction between what is a priori known – before observation – and the empirical knowledge gained from observations. In a Bayesian pattern classifier, the class probabilities $p(label | \theta)$ can be chosen by the user, which are then a priori. Moreover, experience quantified as a priori parameter values can be weighted with empirical observations – using e.g., the Beta- (conjugate prior) and Dirichlet-distributions. The Bayesian approach facilitates a seamless intermixing between expert knowledge in the form of subjective probabilities, and objective observations.

Probabilistic pattern classifiers can be used according to a frequentist or a Bayesian approach.

Uses

The face was automatically detected by special software.

Within medical science, pattern recognition is the basis for computer-aided diagnosis (CAD) systems. CAD describes a procedure that supports the doctor's interpretations and findings.

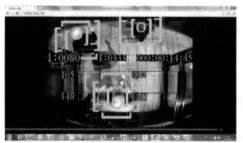

Pattern & Shape Recognition Technology (SRT) in a people counter system.

Other typical applications of pattern recognition techniques are automatic speech recognition, classification of text into several categories (e.g., spam/non-spam email messages), the automatic recognition of handwritten postal codes on postal envelopes, automatic recognition of images of human faces, or handwriting image extraction from medical forms. The last two examples form the subtopic image analysis of pattern recognition that deals with digital images as input to pattern recognition systems.

Optical character recognition is a classic example of the application of a pattern classifier. The method of signing one's name was captured with stylus and overlay starting in 1990. The strokes, speed, relative min, relative max, acceleration and pressure is used to uniquely identify and confirm identity. Banks were first offered this technology, but were content to collect from the FDIC for any bank fraud and did not want to inconvenience customers.

Artificial neural networks (neural net classifiers) and Deep Learning have many real-world applications in image processing, a few examples:

- identification and authentication: e.g., license plate recognition, fingerprint analysis and face detection/verification;

- medical diagnosis: e.g., screening for cervical cancer (Papnet) or breast tumors;

- defence: various navigation and guidance systems, target recognition systems, shape recognition technology etc.

In psychology, pattern recognition (making sense of and identifying objects) is closely related to perception, which explains how the sensory inputs humans receive are made meaningful. Pattern recognition can be thought of in two different ways: the first being template matching and the second being feature detection. A template is a pattern used to produce items of the same proportions. The template-matching hypothesis suggests that incoming stimuli are compared with templates in the long term memory. If there is a match, the stimulus is identified. Feature detection models, such as the Pandemonium system for classifying letters (Selfridge, 1959), suggest that the stimuli are broken down into their component parts for identification. For example, a capital E has three horizontal lines and one vertical line.

Algorithms

Algorithms for pattern recognition depend on the type of label output, on whether learning is supervised or unsupervised, and on whether the algorithm is statistical or non-statistical in nature. Statistical algorithms can further be categorized as generative or discriminative.

Classification Algorithms

Parametric:

- Linear discriminant analysis

- Quadratic discriminant analysis

- Maximum entropy classifier (aka logistic regression, multinomial logistic regression): Note that logistic regression is an algorithm for classification, despite its name. (The name comes from the fact that logistic regression uses an extension of a linear regression model to model the probability of an input being in a particular class.)

Nonparametric:

- Decision trees, decision lists

- Kernel estimation and K-nearest-neighbor algorithms

- Naive Bayes classifier

- Neural networks (multi-layer perceptrons)

- Perceptrons
- Support vector machines
- Gene expression programming

Clustering Algorithms

- Categorical mixture models
- Deep learning methods
- Hierarchical clustering (agglomerative or divisive)
- K-means clustering
- Correlation clustering
- Kernel principal component analysis (Kernel PCA)

Ensemble Learning Algorithms

- Boosting (meta-algorithm)
- Bootstrap aggregating ("bagging")
- Ensemble averaging
- Mixture of experts, hierarchical mixture of experts

General Algorithms for Predicting Arbitrarily-structured (Sets of) Labels

- Bayesian networks
- Markov random fields

Multilinear Subspace Learning Algorithms

Unsupervised:

- Multilinear principal component analysis (MPCA)

Real-valued Sequence Labeling Algorithms

Supervised:

- Kalman filters
- Particle filters

Regression Algorithms

Supervised:

- Gaussian process regression (kriging)

- Linear regression and extensions

- Neural networks and Deep learning methods

Unsupervised:

- Independent component analysis (ICA)

- Principal components analysis (PCA)

Sequence Labeling Algorithms

Supervised:

- Conditional random fields (CRFs)

- Hidden Markov models (HMMs)

- Maximum entropy Markov models (MEMMs)

- Recurrent neural networks

Unsupervised:

- Hidden Markov models (HMMs)

Feature (Machine Learning)

In machine learning and pattern recognition, a feature is an individual measurable property of a phenomenon being observed. Choosing informative, discriminating and independent features is a crucial step for effective algorithms in pattern recognition, classification and regression. Features are usually numeric, but structural features such as strings and graphs are used in syntactic pattern recognition. The concept of "feature" is related to that of explanatory variable used in statistical techniques such as linear regression.

The initial set of raw features can be redundant and too large to be managed. Therefore, a preliminary step in many applications of machine learning and pattern recognition consists of selecting a subset of features, or constructing a new and reduced set of features to facilitate learning, and to improve generalization and interpretability.

Extracting or selecting features is a combination of art and science; developing systems to do so is known as feature engineering. It requires the experimentation of multiple possibilities and the combination of automated techniques with the intuition and knowledge of the domain expert. Automating this process is feature learning, where a machine not only uses features for learning, but learns the features itself.

Classification

A set of numeric features can be conveniently described by a feature vector. An example of reaching a two way classification from a feature vector (related to the perceptron) consists of calculating the scalar product between the feature vector and a vector of weights, comparing the result with a threshold, and deciding the class based on the comparison.

Algorithms for classification from a feature vector include nearest neighbor classification, neural networks, and statistical techniques such as Bayesian approaches.

Examples

In character recognition, features may include histograms counting the number of black pixels along horizontal and vertical directions, number of internal holes, stroke detection and many others.

In speech recognition, features for recognizing phonemes can include noise ratios, length of sounds, relative power, filter matches and many others.

In spam detection algorithms, features may include the presence or absence of certain email headers, the email structure, the language, the frequency of specific terms, the grammatical correctness of the text.

In computer vision, there are a large number of possible features, such as edges and objects.

Feature Learning

In machine learning, feature learning or representation learning is a set of techniques that learn a feature: a transformation of raw data input to a representation that can be effectively exploited in machine learning tasks. This obviates manual feature engineering, which is otherwise necessary, and allows a machine to both learn at a specific task (*using* the features) *and* learn the features themselves.

Feature learning is motivated by the fact that machine learning tasks such as classification often require input that is mathematically and computationally convenient to process. However, real-world data such as images, video, and sensor measurement is usually complex, redundant, and highly variable. Thus, it is necessary to discover useful features or representations from raw data. Traditional hand-crafted features often require expensive human labor and often rely on expert knowledge. Also, they normally do not generalize well. This motivates the design of efficient feature learning techniques, to automate and generalize this.

Feature learning can be divided into two categories: supervised and unsupervised feature learning, analogous to these categories in machine learning generally.

- In supervised feature learning, features are learned with labeled input data. Examples include supervised neural networks, multilayer perceptron, and (supervised) dictionary learning.

- In unsupervised feature learning, features are learned with unlabeled input data. Examples include dictionary learning, independent component analysis, autoencoders, matrix factorization, and various forms of clustering.

Supervised Feature Learning

Supervised feature learning is learning features from labeled data. Several approaches are introduced in the following.

Supervised Dictionary Learning

Dictionary learning is to learn a set (dictionary) of representative elements from the input data such that each data point can be represented as a weighted sum of the representative elements. The dictionary elements and the weights may be found by minimizing the average representation error (over the input data), together with $L1$ regularization on the weights to enable sparsity (i.e., the representation of each data point has only a few nonzero weights).

Supervised dictionary learning exploits both the structure underlying the input data and the labels for optimizing the dictionary elements. For example, a supervised dictionary learning technique was proposed by Mairal et al. in 2009. The authors apply dictionary learning on classification problems by jointly optimizing the dictionary elements, weights for representing data points, and parameters of the classifier based on the input data. In particular, a minimization problem is formulated, where the objective function consists of the classification error, the representation error, an $L1$ regularization on the representing weights for each data point (to enable sparse representation of data), and an $L2$ regularization on the parameters of the classifier.

Neural Networks

Neural networks are used to illustrate a family of learning algorithms via a "network" consisting of multiple layers of inter-connected nodes. It is inspired by the nervous system, where the nodes are viewed as neurons and edges are viewed as synapse. Each edge has an associated weight, and the network defines computational rules that passes input data from the input layer to the output layer. A network function associated with a neural network characterizes the relationship between input and output layers, which is parameterized by the weights. With appropriately defined network functions, various learning tasks can be performed by minimizing a cost function over the network function (weights).

Multilayer neural networks can be used to perform feature learning, since they learn a representation of their input at the hidden layer(s) which is subsequently used for classification or regression at the output layer.

Unsupervised Feature Learning

Unsupervised feature learning is to learn features from unlabeled data. The goal of unsupervised feature learning is often to discover low-dimensional features that captures some structure underlying the high-dimensional input data. When the feature learning is performed in an unsupervised way, it enables a form of semisupervised learning where first, features are learned from an unlabeled dataset, which are then employed to improve performance in a supervised setting with labeled data. Several approaches are introduced in the following.

K-means Clustering

K-means clustering is an approach for vector quantization. In particular, given a set of n vectors, k-means clustering groups them into k clusters (i.e., subsets) in such a way that each vector belongs to the cluster with the closest mean. The problem is computationally NP-hard, and suboptimal greedy algorithms have been developed for k-means clustering.

In feature learning, k-means clustering can be used to group an unlabeled set of inputs into k clusters, and then use the centroids of these clusters to produce features. These features can be produced in several ways. The simplest way is to add k binary features to each sample, where each feature j has value one iff the jth centroid learned by k-means is the closest to the sample under consideration. It is also possible to use the distances to the clusters as features, perhaps after transforming them through a radial basis function (a technique that has used to train RBF networks). Coates and Ng note that certain variants of k-means behave similarly to sparse coding algorithms.

In a comparative evaluation of unsupervised feature learning methods, Coates, Lee and Ng found that k-means clustering with an appropriate transformation outperforms the more recently invented auto-encoders and RBMs on an image classification task. K-means has also been shown to improve performance in the domain of NLP, specifically for named-entity recognition; there, it competes with Brown clustering, as well as with distributed word representations (also known as neural word embeddings).

Principal Component Analysis

Principal component analysis (PCA) is often used for dimension reduction. Given an unlabeled set of n input data vectors, PCA generates p (which is much smaller than the dimension of the input data) right singular vectors corresponding to the p largest singular values of the data matrix, where the kth row of the data matrix is the kth input data vector shifted by the sample mean of the input (i.e., subtracting the sample mean from the data vector). Equivalently, these singular vectors are the eigenvectors corresponding to the p largest eigenvalues of the sample covariance matrix of the input vectors. These p singular vectors are the feature vectors learned from the input data, and they represent directions along which the data has the largest variations.

PCA is a linear feature learning approach since the p singular vectors are linear functions of the data matrix. The singular vectors can be generated via a simple algorithm with p iterations. In the ith iteration, the projection of the data matrix on the $(i\text{-}1)$th eigenvector is subtracted, and the ith singular vector is found as the right singular vector corresponding to the largest singular of the residual data matrix.

PCA has several limitations. First, it assumes that the directions with large variance are of most interest, which may not be the case in many applications. PCA only relies on orthogonal transformations of the original data, and it only exploits the first- and second-order moments of the data, which may not well characterize the distribution of the data. Furthermore, PCA can effectively reduce dimension only when the input data vectors are correlated (which results in a few dominant eigenvalues).

Local Linear Embedding

Local linear embedding (LLE) is a nonlinear unsupervised learning approach for generating low-dimensional neighbor-preserving representations from (unlabeled) high-dimension input. The approach was proposed by Sam T. Roweis and Lawrence K. Saul in 2000.

The general idea of LLE is to reconstruct the original high-dimensional data using lower-dimensional points while maintaining some geometric properties of the neighborhoods in the original data set. LLE consists of two major steps. The first step is for "neighbor-preserving," where each input data point Xi is reconstructed as a weighted sum of K nearest neighboring data points, and the optimal weights are found by minimizing the average squared reconstruction error (i.e., difference between a point and its reconstruction) under the constraint that the weights associated to each point sum up to one. The second step is for "dimension reduction," by looking for vectors in a lower-dimensional space that minimizes the representation error using the optimized weights in the first step. Note that in the first step, the weights are optimized with data being fixed, which can be solved as a least squares problem; while in the second step, lower-dimensional points are optimized with the weights being fixed, which can be solved via sparse eigenvalue decomposition.

The reconstruction weights obtained in the first step captures the "intrinsic geometric properties" of a neighborhood in the input data. It is assumed that original data lie on a smooth lower-dimensional manifold, and the "intrinsic geometric properties" captured by the weights of the original data are expected also on the manifold. This is why the same weights are used in the second step of LLE. Compared with PCA, LLE is more powerful in exploiting the underlying structure of data.

Independent Component Analysis

Independent component analysis (ICA) is a technique for learning a representation of data using a weighted sum of independent non-Gaussian components. The assumption of non-Gaussian is imposed since the weights cannot be uniquely determined when all the components follow Gaussian distribution.

Unsupervised Dictionary Learning

Different from supervised dictionary learning, unsupervised dictionary learning does not utilize the labels of the data and only exploits the structure underlying the data for optimizing the dictionary elements. An example of unsupervised dictionary learning is sparse coding, which aims to learn basis functions (dictionary elements) for data representation from unlabeled input data. Sparse coding can be applied to learn overcomplete dictionaries, where the number of dictionary elements is larger than the dimension of the input data. Aharon et al. proposed an algorithm known as K-SVD for learning from unlabeled input data a dictionary of elements that enables sparse representation of the data.

Multilayer/Deep Architectures

The hierarchical architecture of the neural system inspires deep learning architectures for feature learning by stacking multiple layers of simple learning blocks. These architectures are often designed based on the assumption of distributed representation: observed data is generated by the interactions of many different factors on multiple levels. In a deep learning architecture, the output of each intermediate layer can be viewed as a representation of the original input data. Each level uses the representation produced by previous level as input, and produces new representations as output, which is then fed to higher levels. The input of bottom layer is the raw data, and the output of the final layer is the final low-dimensional feature or representation.

Restricted Boltzmann Machine

Restricted Boltzmann machines (RBMs) are often used as a building block for multilayer learning architectures. An RBM can be represented by an undirected bipartite graph consisting of a group of binary hidden variables, a group of visible variables, and edges connecting the hidden and visible nodes. It is a special case of the more general Boltzmann machines with the constraint of no intra-node connections. Each edge in an RBM is associated with a weight. The weights together with the connections define an energy function, based on which a joint distribution of visible and hidden nodes can be devised. Based on the topology of the RBM, the hidden (visible) variables are independent conditioned on the visible (hidden) variables. Such conditional independence facilitates computations on RBM.

An RBM can be viewed as a single layer architecture for unsupervised feature learning. In particular, the visible variables correspond to input data, and the hidden variables correspond to feature detectors. The weights can be trained by maximizing the probability of visible variables using the contrastive divergence (CD) algorithm by Geoffrey Hinton.

In general, the training of RBM by solving the above maximization problem tends to result in non-sparse representations. The sparse RBM, a modification of the RBM, was proposed to enable sparse representations. The idea is to add a regularization term in the objective function of data likelihood, which penalizes the deviation of the expected hidden variables from a small constant.

Autoencoder

An autoencoder consisting of encoder and decoder is a paradigm for deep learning architectures. An example is provided by Hinton and Salakhutdinov where the encoder uses raw data (e.g., image) as input and produces feature or representation as output, and the decoder uses the extracted feature from the encoder as input and reconstructs the original input raw data as output. The encoder and decoder are constructed by stacking multiple layers of RBMs. The parameters involved in the architecture were originally trained in a greedy layer-by-layer manner: after one layer of feature detectors is learned, they are fed to upper layers as visible variables for training the corresponding RBM. Current approaches typically apply end-to-end training with stochastic gradient descent methods. Training can be repeated until some stopping criteria is satisfied.

Grammar Induction

Grammar induction, also known as grammatical inference or syntactic pattern recognition, refers to the process in machine learning of learning a formal grammar (usually as a collection of *rewrite rules* or *productions* or alternatively as a finite state machine or automaton of some kind) from a set of observations, thus constructing a model which accounts for the characteristics of the observed objects. More generally, grammatical inference is that branch of machine learning where the instance space consists of discrete combinatorial objects such as strings, trees and graphs.

Grammar Classes

Grammatical inference has often been very focused on the problem of learning finite state machines of various types. since there have been efficient algorithms for this problem since the 1980s.

More recently these approaches have been extended to the problem of inference of context-free grammars and richer formalisms, such as multiple context-free grammars and parallel multiple context-free grammars. Other classes of grammars for which grammatical inference has been studied are contextual grammars and pattern languages.

Learning Models

The simplest form of learning is where the learning algorithm merely receives a set of examples drawn from the language in question, but other learning models have been studied. One frequently studied alternative is the case where the learner can ask membership queries as in the exact query learning model or minimally adequate teacher model introduced by Angluin.

Methodologies

There is a wide variety of methods for grammatical inference. Two of the classic sources are Fu (1977) and Fu (1982). Duda, Hart & Stork (2001) also devote a brief section to the problem, and cite a number of references. The basic trial-and-error method they present is discussed below. For approaches to infer subclasses of regular languages in particular. A more recent textbook is de la Higuera (2010), which covers the theory of grammatical inference of regular languages and finite state automata. D'Ulizia, Ferri and Grifoni provide a survey that explores grammatical inference methods for natural languages.

Grammatical Inference by Trial-and-error

The method proposed in Section 8.7 of Duda, Hart & Stork (2001) suggests successively guessing grammar rules (productions) and testing them against positive and negative observations. The rule set is expanded so as to be able to generate each positive example, but if a given rule set also generates a negative example, it must be discarded. This particular approach can be characterized as "hypothesis testing" and bears some similarity to Mitchel's version space algorithm. The Duda, Hart & Stork (2001) text provide a simple example which nicely illustrates the process, but the feasibility of such an unguided trial-and-error approach for more substantial problems is dubious.

Grammatical Inference by Genetic Algorithms

Grammatical induction using evolutionary algorithms is the process of evolving a representation of the grammar of a target language through some evolutionary process. Formal grammars can easily be represented as tree structures of production rules that can be subjected to evolutionary operators. Algorithms of this sort stem from the genetic programming paradigm pioneered by John Koza. Other early work on simple formal languages used the binary string representation of genetic algorithms, but the inherently hierarchical structure of grammars couched in the EBNF language made trees a more flexible approach.

Koza represented Lisp programs as trees. He was able to find analogues to the genetic operators within the standard set of tree operators. For example, swapping sub-trees is equivalent to the corresponding process of genetic crossover, where sub-strings of a genetic code are transplanted into an individual of the next generation. Fitness is measured by scoring the output from the functions of the Lisp code. Similar analogues between the tree structured lisp representation and the representation of grammars as trees, made the application of genetic programming techniques possible for grammar induction.

In the case of grammar induction, the transplantation of sub-trees corresponds to the swapping of production rules that enable the parsing of phrases from some language. The fitness operator for the grammar is based upon some measure of how well it performed in parsing some group of sentences from the target language. In a tree representation of a grammar, a terminal symbol of a production rule corresponds to a leaf node of the tree. Its parent nodes corresponds to a non-terminal symbol (e.g. a noun phrase or a verb phrase) in the rule set. Ultimately, the root node might correspond to a sentence non-terminal.

Grammatical Inference by Greedy Algorithms

Like all greedy algorithms, greedy grammar inference algorithms make, in iterative manner, decisions that seem to be the best at that stage. The decisions made usually deal with things like the creation of new rules, the removal of existing rules, the choice of a rule to be applied or the merging of some existing rules. Because there are several ways to define 'the stage' and 'the best', there are also several greedy grammar inference algorithms.

These context-free grammar generating algorithms make the decision after every read symbol:

- Lempel-Ziv-Welch algorithm creates a context-free grammar in a deterministic way such that it is necessary to store only the start rule of the generated grammar.

- Sequitur and its modifications.

These context-free grammar generating algorithms first read the whole given symbol-sequence and then start to make decisions:

- Byte pair encoding and its optimizations.

Distributional Learning

A more recent approach is based on distributional learning. Algorithms using these approaches have been applied to learning context-free grammars and mildly context-sensitive languages and have been proven to be correct and efficient for large subclasses of these grammars.

Learning of Pattern Languages

Angluin defines a *pattern* to be "a string of constant symbols from Σ and variable symbols from a disjoint set". The language of such a pattern is the set of all its nonempty ground instances i.e. all strings resulting from consistent replacement of its variable symbols by nonempty strings of constant symbols. A pattern is called descriptive for a finite input set of strings if its language is minimal (with respect to set inclusion) among all pattern languages subsuming the input set.

Angluin gives a polynomial algorithm to compute, for a given input string set, all descriptive patterns in one variable x. To this end, she builds an automaton representing all possibly relevant patterns; using sophisticated arguments about word lengths, which rely on x being the only variable, the state count can be drastically reduced.

Erlebach et al. give a more efficient version of Angluin's pattern learning algorithm, as well as a parallelized version.

Arimura et al. show that a language class obtained from limited unions of patterns can be learned in polynomial time.

Pattern Theory

Pattern theory, formulated by Ulf Grenander, is a mathematical formalism to describe knowledge of the world as patterns. It differs from other approaches to artificial intelligence in that it does not begin by prescribing algorithms and machinery to recognize and classify patterns; rather, it prescribes a vocabulary to articulate and recast the pattern concepts in precise language.

In addition to the new algebraic vocabulary, its statistical approach was novel in its aim to:

- Identify the hidden variables of a data set using real world data rather than artificial stimuli, which was commonplace at the time.

- Formulate prior distributions for hidden variables and models for the observed variables that form the vertices of a Gibbs-like graph.

- Study the randomness and variability of these graphs.

- Create the basic classes of stochastic models applied by listing the deformations of the patterns.

- Synthesize (sample) from the models, not just analyze signals with it.

Broad in its mathematical coverage, pattern theory spans algebra and statistics, as well as local topological and global entropic properties.

Applications

The principle of grammar induction has been applied to other aspects of natural language processing, and has been applied (among many other problems) to morpheme analysis, and place name derivations. Grammar induction has also been used for lossless data compression and statistical inference via minimum message length (MML) and minimum description length (MDL) principles.

Dimensionality Reduction

In machine learning and statistics, dimensionality reduction or dimension reduction is the process of reducing the number of random variables under consideration, via obtaining a set of principal variables. It can be divided into feature selection and feature extraction.

Feature Selection

Feature selection approaches try to find a subset of the original variables (also called features or attributes). There are three strategies; *filter* (e.g. information gain) and *wrapper* (e.g. search guided by accuracy) approaches, and *embedded* (features are selected to add or be removed while building the model based on the prediction errors).

In some cases, data analysis such as regression or classification can be done in the reduced space more accurately than in the original space.

Feature Extraction

Feature extraction transforms the data in the high-dimensional space to a space of fewer dimensions. The data transformation may be linear, as in principal component analysis (PCA), but many nonlinear dimensionality reduction techniques also exist. For multidimensional data, tensor representation can be used in dimensionality reduction through multilinear subspace learning.

Principal Component Analysis (PCA)

The main linear technique for dimensionality reduction, principal component analysis, performs a linear mapping of the data to a lower-dimensional space in such a way that the variance of the data in the low-dimensional representation is maximized. In practice, the covariance (and sometimes the correlation) matrix of the data is constructed and the eigen vectors on this matrix are computed. The eigen vectors that correspond to the largest eigenvalues (the principal components) can now be used to reconstruct a large fraction of the variance of the original data. Moreover, the first few eigen vectors can often be interpreted in terms of the large-scale physical behavior of the system. The original space (with dimension of the number of points) has been reduced (with data loss, but hopefully retaining the most important variance) to the space spanned by a few eigenvectors.

Kernel PCA

Principal component analysis can be employed in a nonlinear way by means of the kernel trick. The resulting technique is capable of constructing nonlinear mappings that maximize the variance in the data. The resulting technique is entitled kernel PCA.

Graph-based Kernel PCA

Other prominent nonlinear techniques include manifold learning techniques such as Isomap, locally linear embedding (LLE), Hessian LLE, Laplacian eigenmaps, and local tangent space alignment (LTSA). These techniques construct a low-dimensional data representation using a cost function that retains local properties of the data, and can be viewed as defining a graph-based kernel for Kernel PCA.

More recently, techniques have been proposed that, instead of defining a fixed kernel, try to learn the kernel using semidefinite programming. The most prominent example of such a technique is maximum variance unfolding (MVU). The central idea of MVU is to exactly preserve all pairwise distances between nearest neighbors (in the inner product space), while maximizing the distances between points that are not nearest neighbors.

An alternative approach to neighborhood preservation is through the minimization of a cost function that measures differences between distances in the input and output spaces. Important examples of such techniques include: classical multidimensional scaling, which is identical to PCA; Isomap, which uses geodesic distances in the data space; diffusion maps, which use diffusion distances in the data space; t-distributed stochastic neighbor embedding (t-SNE), which minimizes the divergence between distributions over pairs of points; and curvilinear component analysis.

A different approach to nonlinear dimensionality reduction is through the use of autoencoders, a special kind of feed-forward neural networks with a bottle-neck hidden layer. The training of deep encoders is typically performed using a greedy layer-wise pre-training (e.g., using a stack of restricted Boltzmann machines) that is followed by a finetuning stage based on backpropagation.

Linear Discriminant Analysis (LDA)

Linear discriminant analysis (LDA) is a generalization of fisher's linear discriminant, a method used in statistics, pattern recognition and machine learning to find a linear combination of features that characterizes or separates two or more classes of objects or events.

Generalized Discriminant Analysis (GDA)

GDA deals with nonlinear discriminant analysis using kernel function operator. The underlying theory is close to the support vector machines (SVM) insofar as the GDA method provides a mapping of the input vectors into high-dimensional feature space. Similar to LDA, the objective of GDA is to find a projection for the features into a lower dimensional space by maximizing the ratio of between-class scatter to within-class scatter.

Dimension Reduction

For high-dimensional datasets (i.e. with number of dimensions more than 10), dimension reduction is usually performed prior to applying a K-nearest neighbors algorithm (k-NN) in order to avoid the effects of the curse of dimensionality.

Feature extraction and dimension reduction can be combined in one step using principal component analysis (PCA), linear discriminant analysis (LDA), or canonical correlation analysis (CCA) techniques as a pre-processing step followed by clustering by K-NN on feature vectors in reduced-dimension space. In machine learning this process is also called low-dimensional embedding.

For very-high-dimensional datasets (e.g. when performing similarity search on live video streams, DNA data or high-dimensional time series) running a fast approximate K-NN search using locality sensitive hashing, random projection, "sketches" or other high-dimensional similarity search techniques from the VLDB toolbox might be the only feasible option.

Advantages of Dimensionality Reduction

1. It reduces the time and storage space required.

2. Removal of multi-collinearity improves the performance of the machine learning model.

3. It becomes easier to visualize the data when reduced to very low dimensions such as 2D or 3D.

Applications

A dimensionality reduction technique that is sometimes used in neuroscience is maximally informative dimensions, which finds a lower-dimensional representation of a dataset such that as much information as possible about the original data is preserved.

Gradient Boosting

Gradient boosting is a machine learning technique for regression and classification problems, which produces a prediction model in the form of an ensemble of weak prediction models, typically decision trees. It builds the model in a stage-wise fashion like other boosting methods do, and it generalizes them by allowing optimization of an arbitrary differentiable loss function.

The idea of gradient boosting originated in the observation by Leo Breiman that boosting can be interpreted as an optimization algorithm on a suitable cost function. Explicit regression gradient boosting algorithms were subsequently developed by Jerome H. Friedman simultaneously with the more general functional gradient boosting perspective of Llew Mason, Jonathan Baxter, Peter Bartlett and Marcus Frean. The latter two papers introduced the abstract view of boosting algorithms as iterative *functional gradient descent* algorithms. That is, algorithms that optimize a cost *function* over function space by iteratively choosing a function (weak hypothesis) that points in the negative gradient direction. This functional gradient view of boosting has led to the development of boosting algorithms in many areas of machine learning and statistics beyond regression and classification.

Informal Introduction

Like other boosting methods, gradient boosting combines weak "learners" into a single strong learner, in an iterative fashion. It is easiest to explain in the least-squares regression setting, where the goal is to "teach" a model F to predict values in the form $\hat{y} = F(x)$, by minimizing the mean squared error $(\hat{y} - y)^2$ to the true values y (averaged over some training set).

At each stage $1 \le m \le M$ of gradient boosting, it may be assumed that there is some imperfect model F_m (at the outset, a very weak model that just predicts the mean y in the training set could be used). The gradient boosting algorithm does not change F_m in any way; instead, it improves on it by constructing a new model that adds an estimator h to provide a better model $F_{m+1}(x) = F_m(x) + h(x)$. The question is now, how to find h? The gradient boosting solution starts with the observation that a perfect h would imply

$$F_{m+1}(x) = F_m(x) + h(x) = y$$

or, equivalently,

$$h(x) = y - F_m(x).$$

Therefore, gradient boosting will fit h to the *residual* $y - F_m(x)$. Like in other boosting variants, each F_{m+1} learns to correct its predecessor F_m. A generalization of this idea to other loss functions than squared error (and to classification and ranking problems) follows from the observation that residuals $y - F(x)$ are the negative gradients of the squared error loss function $\frac{1}{2}(y - F(x))^2$. So, gradient boosting is a gradient descent algorithm; and generalizing it entails "plugging in" a different loss and its gradient.

Algorithm

In many supervised learning problems one has an output variable y and a vector of input variables x connected together via a joint probability distribution $P(x, y)$. Using a training set $\{(x_1, y_1), \ldots, (x_n, y_n)\}$ of known values of x and corresponding values of y, the goal is to find an approximation $\hat{F}(x)$ to a function $F^*(x)$ that minimizes the expected value of some specified loss function $L(y, F(x))$:

$$\hat{F} = \arg\min_F \mathbb{E}_{x,y}[L(y, F(x))] \quad .$$

The gradient boosting method assumes a real-valued y and seeks an approximation $\hat{F}(x)$ in the form of a weighted sum of functions $h_i(x)$ from some class H, called base (or weak) learners:

$$F(x) = \sum_{i=1}^{M} \gamma_i h_i(x) + const.$$

In accordance with the empirical risk minimization principle, the method tries to find an approximation $\hat{F}(x)$ that minimizes the average value of the loss function on the training set. It does so by starting with a model, consisting of a constant function $F_0(x)$, and incrementally expanding it in a greedy fashion:

$$F_0(x) = \arg\min_\gamma \sum_{i=1}^{n} L(y_i, \gamma),$$

$$F_m(x) = F_{m-1}(x) + \arg\min_{h \in \mathcal{H}} \sum_{i=1}^{n} L(y_i, F_{m-1}(x_i) + h(x_i))$$

,

where $h \in H$ is a base learner function.

Unfortunately, choosing the best function h at each step for an arbitrary loss function L is a computationally infeasible optimization problem in general. Therefore, we will restrict to a simplification.

The idea is to apply a steepest descent step to this minimization problem. If we considered the continuous case, i.e. H the set of arbitrary differentiable functions on $\mathbb{R}$, we would update the model in accordance with the following equations

$$F_m(x) = F_{m-1}(x) - \gamma_m \sum_{i=1}^{n} \nabla_{F_{m-1}} L(y_i, F_{m-1}(x_i)),$$

$$\gamma_m = \arg\min_{\gamma} \sum_{i=1}^{n} L\left(y_i, F_{m-1}(x_i) - \gamma \frac{\partial L(y_i, F_{m-1}(x_i))}{\partial F_{m-1}(x_i)} \right),$$

where the derivatives are taken with respect to the functions F_i for $i \in \{1,..,m\}$. In the discrete case however, i.e. the set H is finite, we will choose the candidate function h closest to the gradient of L for which the coefficient γ may then be calculated with the aid of line search the above equations. Note that this approach is a heuristic and will therefore not yield an exact solution to the given problem, yet a satisfactory approximation.

In pseudocode, the generic gradient boosting method is:

Input: training set $\{(x_i, y_i)\}_{i=1}^{n}$, a differentiable loss function $L(y, F(x))$ number of iterations M.

Algorithm:

1. Initialize model with a constant value:

$$F_0(x) = \arg\min_{\gamma} \sum_{i=1}^{n} L(y_i, \gamma).$$

2. For $m = 1$ to M:

    1. Compute so-called *pseudo-residuals*:

    $$r_{im} = -\left[\frac{\partial L(y_i, F(x_i))}{\partial F(x_i)} \right]_{F(x)=F_{m-1}(x)} \quad \text{for } i = 1,\ldots,n.$$

    2. Fit a base learner $h_m(x)$ to pseudo-residuals, i.e. train it using the training set $\{(x_i, r_{im})\}_{i=1}^{n}$.

    3. Compute multiplier γ_m by solving the following one-dimensional optimization problem:

    $$\gamma_m = \arg\min_{\gamma} \sum_{i=1}^{n} L\left(y_i, F_{m-1}(x_i) + \gamma h_m(x_i) \right).$$

4. Update the model:

$$F_m(x) = F_{m-1}(x) + \gamma_m h_m(x).$$

3. Output $F_M(x)$.

Gradient Tree Boosting

Gradient boosting is typically used with decision trees (especially CART trees) of a fixed size as base learners. For this special case Friedman proposes a modification to gradient boosting method which improves the quality of fit of each base learner.

Generic gradient boosting at the m-th step would fit a decision tree $h_m(x)$ to pseudo-residuals. Let J_m be the number of its leaves. The tree partitions the input space into J_m disjoint regions $R_{1m}, \ldots, R_{J_m m}$ and predicts a constant value in each region. Using the indicator notation, the output of $h_m(x)$ for input x can be written as the sum:

$$h_m(x) = \sum_{j=1}^{J_m} b_{jm} I(x \in R_{jm})$$

where b_{jm} is the value predicted in the region R_{jm}.

Then the coefficients b_{jm} are multiplied by some value γ_m, chosen using line search so as to minimize the loss function, and the model is updated as follows:

$$F_m(x) = F_{m-1}(x) + \gamma_m h_m(x), \quad \gamma_m = \arg\min_{\gamma} \sum_{i=1}^{n} L(y_i, F_{m-1}(x_i) + \gamma h_m(x_i)).$$

Friedman proposes to modify this algorithm so that it chooses a separate optimal value γ_{jm} for each of the tree's regions, instead of a single γ_{jm} for the whole tree. He calls the modified algorithm "TreeBoost". The coefficients b_{jm} from the tree-fitting procedure can be then simply discarded and the model update rule becomes:

$$F_m(x) = F_{m-1}(x) + \sum_{j=1}^{J_m} \gamma_{jm} I(x \in R_{jm}), \quad \gamma_{jm} = \arg\min_{\gamma} \sum_{x_i \in R_{jm}} L(y_i, F_{m-1}(x_i) + \gamma).$$

Size of Trees

J, the number of terminal nodes in trees, is the method's parameter which can be adjusted for a data set at hand. It controls the maximum allowed level of interaction between variables in the model. With $J = 2$ (decision stumps), no interaction between variables is allowed. With $J = 3$ the model may include effects of the interaction between up to two variables, and so on.

Hastie et al. comment that typically $4 \le J \le 8$ work well for boosting and results are fairly insensitive to the choice of J in this range, $J = 2$ is insufficient for many applications, and $J > 10$ is unlikely to be required.

Regularization

Fitting the training set too closely can lead to degradation of the model's generalization ability. Several so-called regularization techniques reduce this overfitting effect by constraining the fitting procedure.

One natural regularization parameter is the number of gradient boosting iterations M (i.e. the number of trees in the model when the base learner is a decision tree). Increasing M reduces the error on training set, but setting it too high may lead to overfitting. An optimal value of M is often selected by monitoring prediction error on a separate validation data set. Besides controlling M, several other regularization techniques are used.

Shrinkage

An important part of gradient boosting method is regularization by shrinkage which consists in modifying the update rule as follows:

$$F_m(x) = F_{m-1}(x) + v \cdot \gamma_m h_m(x), \quad 0 < v \le 1,$$

where parameter v is called the "learning rate".

Empirically it has been found that using small learning rates (such as $v < 0.1$) yields dramatic improvements in model's generalization ability over gradient boosting without shrinking ($v = 1$). However, it comes at the price of increasing computational time both during training and querying: lower learning rate requires more iterations.

Stochastic Gradient Boosting

Soon after the introduction of gradient boosting Friedman proposed a minor modification to the algorithm, motivated by Breiman's bagging method. Specifically, he proposed that at each iteration of the algorithm, a base learner should be fit on a subsample of the training set drawn at random without replacement. Friedman observed a substantial improvement in gradient boosting's accuracy with this modification.

Subsample size is some constant fraction f of the size of the training set. When $f = 1$, the algorithm is deterministic and identical to the one described above. Smaller values of f introduce randomness into the algorithm and help prevent overfitting, acting as a kind of regularization. The algorithm also becomes faster, because regression trees have to be fit to smaller datasets at each iteration. Friedman obtained that $0.5 \le f \le 0.8$ leads to good results for small and moderate sized training sets. Therefore, f is typically set to 0.5, meaning that one half of the training set is used to build each base learner.

Also, like in bagging, subsampling allows one to define an out-of-bag error of the prediction performance improvement by evaluating predictions on those observations which were not used in the building of the next base learner. Out-of-bag estimates help avoid the need for an independent validation dataset, but often underestimate actual performance improvement and the optimal number of iterations.

Number of Observations in Leaves

Gradient tree boosting implementations often also use regularization by limiting the minimum number of observations in trees' terminal nodes (this parameter is called n.minobsinnode in the R gbm package). It is used in the tree building process by ignoring any splits that lead to nodes containing fewer than this number of training set instances.

Imposing this limit helps to reduce variance in predictions at leaves.

Penalize Complexity of Tree

Another useful regularization techniques for gradient boosted trees is to penalize model complexity of the learned model. The model complexity can be defined as the proportional number of leaves in the learned trees. The joint optimization of loss and model complexity corresponds to a post-pruning algorithm to remove branches that fail to reduce the loss by a threshold. Other kinds of regularization such as an ℓ_2 penalty on the leaf values can also be added to avoid overfitting.

Usage

Gradient boosting can be used in the field of learning to rank. The commercial web search engines Yahoo and Yandex use variants of gradient boosting in their machine-learned ranking engines.

Names

The method goes by a variety of names. Friedman introduced his regression technique as a "Gradient Boosting Machine" (GBM). Mason, Baxter et. el. described the generalized abstract class of algorithms as "functional gradient boosting".

A popular open-source implementation for R calls it "Generalized Boosting Model". Commercial implementations from Salford Systems use the names "Multiple Additive Regression Trees" (MART) and TreeNet, both trademarked.

References

- Hyvärinen, Aapo; Oja, Erkki (2000). "Independent Component Analysis: Algorithms and Applications". Neural Networks. 13 (4): 411–430. PMID 10946390. doi:10.1016/s0893-6080(00)00026-5

- Isabelle Guyon Clopinet, André Elisseeff (2003). An Introduction to Variable and Feature Selection. The Journal of Machine Learning Research, Vol. 3, 1157-1182

- Richard O. Duda, Peter E. Hart, David G. Stork (2001). Pattern classification (2nd ed.). Wiley, New York. ISBN 0-471-05669-3

- Carvalko, J.R., Preston K. (1972). "On Determining Optimum Simple Golay Marking Transforms for Binary Image Processing". IEEE Transactions on Computers. 21: 1430–33. doi:10.1109/T-C.1972.223519

- Clark and Eyraud (2007) Journal of Machine Learning Research; Ryo Yoshinaka (2011) Theoretical Computer Science

- "A-level Psychology Attention Revision - Pattern recognition | S-cool, the revision website". S-cool.co.uk. Retrieved 2012-09-17

- Mineichi Kudo; Jack Sklansky (2000). "Comparison of algorithms that select features for pattern classifiers". Pattern Recognition. 33 (1): 25–41. doi:10.1016/S0031-3203(99)00041-2

- Zahorian, Stephen A.; Hu, Hongbing (2011). "Nonlinear Dimensionality Reduction Methods for Use with Automatic Speech Recognition". Speech Technologies. ISBN 978-953-307-996-7. doi:10.5772/16863

- Bengio, Yoshua (2009). "Learning Deep Architectures for AI". Foundations and Trends in Machine Learning. 2 (1): 1–127. doi:10.1561/2200000006

- Dana Angluin (1980). "Finding Patterns Common to a Set of Strings" (PDF). Journal of Computer and System Sciences. 21: 46–62. doi:10.1016/0022-0000(80)90041-0

- Egmont-Petersen, M., de Ridder, D., Handels, H. (2002). "Image processing with neural networks - a review". Pattern Recognition. 35 (10): 2279–2301. doi:10.1016/S0031-3203(01)00178-9

- Hastie, T.; Tibshirani, R.; Friedman, J. H. (2009). "10. Boosting and Additive Trees". The Elements of Statistical Learning (2nd ed.). New York: Springer. pp. 337–384. ISBN 0-387-84857-6

- Lakshmi Padmaja, Dhyaram; Vishnuvardhan, B (18 August 2016). "Comparative Study of Feature Subset Selection Methods for Dimensionality Reduction on Scientific Data": 31–34. doi:10.1109/IACC.2016.16. Retrieved 7 October 2016

- Dana Angluin (1987). "Learning Regular Sets from Queries and Counter-Examples" (PDF). Information and Control. 75: 87–106. doi:10.1016/0890-5401(87)90052-6. Archived from the original (PDF) on 2013-12-02

Supervised Learning and Unsupervised Learning: Machine Learning Tools

Supervised learning infers tasks from labeled data while unsupervised learning uses unlabeled data for computing a function. Semi-supervised learning, structured prediction, bias–variance tradeoff, competitive learning and autoencoder are some important topics related to the subject matter. The major components of supervised and unsupervised learning are discussed in this chapter.

Supervised Learning

Supervised learning is the machine learning task of inferring a function from *labeled training data*. The training data consist of a set of *training examples*. In supervised learning, each example is a *pair* consisting of an input object (typically a vector) and a desired output value (also called the *supervisory signal*). A supervised learning algorithm analyzes the training data and produces an inferred function, which can be used for mapping new examples. An optimal scenario will allow for the algorithm to correctly determine the class labels for unseen instances. This requires the learning algorithm to generalize from the training data to unseen situations in a "reasonable" way.

The parallel task in human and animal psychology is often referred to as concept learning.

In order to solve a given problem of supervised learning, one has to perform the following steps:

1. Determine the type of training examples. Before doing anything else, the user should decide what kind of data is to be used as a training set. In the case of handwriting analysis, for example, this might be a single handwritten character, an entire handwritten word, or an entire line of handwriting.

2. Gather a training set. The training set needs to be representative of the real-world use of the function. Thus, a set of input objects is gathered and corresponding outputs are also gathered, either from human experts or from measurements.

3. Determine the input feature representation of the learned function. The accuracy of the learned function depends strongly on how the input object is represented. Typically, the input object is transformed into a feature vector, which contains a number of features that are descriptive of the object. The number of features should not be too large, because of the curse of dimensionality; but should contain enough information to accurately predict the output.

4. Determine the structure of the learned function and corresponding learning algorithm. For example, the engineer may choose to use support vector machines or decision trees.

5. Complete the design. Run the learning algorithm on the gathered training set. Some supervised learning algorithms require the user to determine certain control parameters. These parameters may be adjusted by optimizing performance on a subset (called a *validation set*) of the training set, or via cross-validation.

6. Evaluate the accuracy of the learned function. After parameter adjustment and learning, the performance of the resulting function should be measured on a test set that is separate from the training set.

A wide range of supervised learning algorithms are available, each with its strengths and weaknesses. There is no single learning algorithm that works best on all supervised learning problems.

There are four major issues to consider in supervised learning:

Bias-variance Tradeoff

A first issue is the tradeoff between *bias* and *variance*. Imagine that we have available several different, but equally good, training data sets. A learning algorithm is biased for a particular input x if, when trained on each of these data sets, it is systematically incorrect when predicting the correct output for x. A learning algorithm has high variance for a particular input x if it predicts different output values when trained on different training sets. The prediction error of a learned classifier is related to the sum of the bias and the variance of the learning algorithm. Generally, there is a tradeoff between bias and variance. A learning algorithm with low bias must be "flexible" so that it can fit the data well. But if the learning algorithm is too flexible, it will fit each training data set differently, and hence have high variance. A key aspect of many supervised learning methods is that they are able to adjust this tradeoff between bias and variance (either automatically or by providing a bias/variance parameter that the user can adjust).

Function Complexity and Amount of Training Data

The second issue is the amount of training data available relative to the complexity of the "true" function (classifier or regression function). If the true function is simple, then an "inflexible" learning algorithm with high bias and low variance will be able to learn it from a small amount of data. But if the true function is highly complex (e.g., because it involves complex interactions among many different input features and behaves differently in different parts of the input space), then the function will only be learnable from a very large amount of training data and using a "flexible" learning algorithm with low bias and high variance.

Dimensionality of the Input Space

A third issue is the dimensionality of the input space. If the input feature vectors have very high dimension, the learning problem can be difficult even if the true function only depends on a small number of those features. This is because the many "extra" dimensions can confuse the learning algorithm and cause it to have high variance. Hence, high input dimensionality typically requires tuning the classifier to have low variance and high bias. In practice, if the engineer can manually remove irrelevant features from the input data, this is likely to improve the accuracy of the learned function. In addition, there are many algorithms for feature selection that seek to identify the relevant features and discard the irrelevant ones. This is an instance of the more general strategy of dimensionality reduction, which seeks to map the input data into a lower-dimensional space prior to running the supervised learning algorithm.

Noise in the Output Values

A fourth issue is the degree of noise in the desired output values (the supervisory target variables). If the desired output values are often incorrect (because of human error or sensor errors), then the learning algorithm should not attempt to find a function that exactly matches the training examples. Attempting to fit the data too carefully leads to overfitting. You can overfit even when there are no measurement errors (stochastic noise) if the function you are trying to learn is too complex for your learning model. In such a situation that part of the target function that cannot be modeled "corrupts" your training data - this phenomenon has been called deterministic noise. When either type of noise is present, it is better to go with a higher bias, lower variance estimator.

In practice, there are several approaches to alleviate noise in the output values such as early stopping to prevent overfitting as well as detecting and removing the noisy training examples prior to training the supervised learning algorithm. There are several algorithms that identify noisy training examples and removing the suspected noisy training examples prior to training has decreased generalization error with statistical significance.

Other Factors to Consider

Other factors to consider when choosing and applying a learning algorithm include the following:

1. Heterogeneity of the data. If the feature vectors include features of many different kinds (discrete, discrete ordered, counts, continuous values), some algorithms are easier to apply than others. Many algorithms, including Support Vector Machines, linear regression, logistic regression, neural networks, and nearest neighbor methods, require that the input features be numerical and scaled to similar ranges (e.g., to the [-1,1] interval). Methods that employ a distance function, such as nearest neighbor methods and support vector machines with Gaussian kernels, are particularly sensitive to this. An advantage of decision trees is that they easily handle heterogeneous data.

2. Redundancy in the data. If the input features contain redundant information (e.g., highly correlated features), some learning algorithms (e.g., linear regression, logistic regression, and distance based methods) will perform poorly because of numerical instabilities. These problems can often be solved by imposing some form of regularization.

3. Presence of interactions and non-linearities. If each of the features makes an independent contribution to the output, then algorithms based on linear functions (e.g., linear regression, logistic regression, Support Vector Machines, naive Bayes) and distance functions (e.g., nearest neighbor methods, support vector machines with Gaussian kernels) generally perform well. However, if there are complex interactions among features, then algorithms such as decision trees and neural networks work better, because they are specifically designed to discover these interactions. Linear methods can also be applied, but the engineer must manually specify the interactions when using them.

When considering a new application, the engineer can compare multiple learning algorithms and experimentally determine which one works best on the problem at hand. Tuning the performance of a learning algorithm can be very time-consuming. Given fixed resources, it is often better to spend more time collecting additional training data and more informative features than it is to spend extra time tuning the learning algorithms.

The most widely used learning algorithms are Support Vector Machines, linear regression, logistic regression, naive Bayes, linear discriminant analysis, decision trees, k-nearest neighbor algorithm, and Neural Networks (Multilayer perceptron).

How Supervised Learning Algorithms Work

Given a set of N training examples of the form $\{(x_1, y_1), ..., (x_N, y_N)\}$ such that x_i is the feature vector of the i-th example and y_i is its label (i.e., class), a learning algorithm seeks a function $g : X \rightarrow Y$, where X is the input space and Y is the output space. The function g is an element of some space of possible functions G, usually called the *hypothesis space*. It is sometimes convenient to represent g using a scoring function $f : X \times Y \rightarrow \mathbb{R}$ such that g is defined as returning the y value that gives the highest score: $g(x) = \arg\max_y f(x, y)$. Let F denote the space of scoring functions.

Although G and F can be any space of functions, many learning algorithms are probabilistic models where g takes the form of a conditional probability model $g(x) = P(y | x)$, or f takes the form of a joint probability model $f(x, y) = P(x, y)$. For example, naive Bayes and linear discriminant analysis are joint probability models, whereas logistic regression is a conditional probability model.

There are two basic approaches to choosing f or g: empirical risk minimization and structural risk minimization. Empirical risk minimization seeks the function that best fits the training data. Structural risk minimize includes a *penalty function* that controls the bias/variance tradeoff.

In both cases, it is assumed that the training set consists of a sample of independent and identically distributed pairs, (x_i, y_i). In order to measure how well a function fits the training data, a loss function $L : Y \times Y \rightarrow \mathbb{R}^{\geq 0}$ is defined. For training example (x_i, y_i), the loss of predicting the value $\hat{y}$ is $L(y_i, \hat{y})$.

The *risk* $R(g)$ of function g is defined as the expected loss of g. This can be estimated from the training data as

$$R_{emp}(g) = \frac{1}{N}\sum_i L(y_i, g(x_i))$$

Empirical Risk Minimization

In empirical risk minimization, the supervised learning algorithm seeks the function g that minimizes $R(g)$. Hence, a supervised learning algorithm can be constructed by applying an optimization algorithm to find g.

When g is a conditional probability distribution $P(y | x)$ and the loss function is the negative log likelihood: $L(y, \hat{y}) = -\log P(y | x)$, then empirical risk minimization is equivalent to maximum likelihood estimation.

When G contains many candidate functions or the training set is not sufficiently large, empirical risk minimization leads to high variance and poor generalization. The learning algorithm is able to memorize the training examples without generalizing well. This is called overfitting.

Structural Risk Minimization

Structural risk minimization seeks to prevent overfitting by incorporating a regularization penalty into the optimization. The regularization penalty can be viewed as implementing a form of Occam's razor that prefers simpler functions over more complex ones.

A wide variety of penalties have been employed that correspond to different definitions of complexity. For example, consider the case where the function g is a linear function of the form

$$g(x) = \sum_{j\,1} \beta_j x_j.$$

A popular regularization penalty is $\sum_j \beta_j^2$, which is the squared Euclidean norm of the weights, also known as the L_2 norm. Other norms include the L_1 norm, $\sum_j |\beta_j|$, and the L_0 norm, which is the number of non-zero β_j s. The penalty will be denoted by $C(g)$.

The supervised learning optimization problem is to find the function g that minimizes

$$J(g) = R_{emp}(g) + \lambda C(g).$$

The parameter λ controls the bias-variance tradeoff. When $\lambda = 0$, this gives empirical risk minimization with low bias and high variance. When λ is large, the learning algorithm will have high bias and low variance. The value of λ can be chosen empirically via cross validation.

The complexity penalty has a Bayesian interpretation as the negative log prior probability of g, $-\log P(g)$, in which case $J(g)$ is the posterior probabability of g.

Generative Training

The training methods described above are *discriminative training* methods, because they seek to find a function g that discriminates well between the different output values. For the special case where $f(x, y) = P(x, y)$ is a joint probability distribution and the loss function is the negative log likelihood $-\sum_i \log P(x_i, y_i)$ a risk minimization algorithm is said to perform *generative training*, because f can be regarded as a generative model that explains how the data were generated. Generative training algorithms are often simpler and more computationally efficient than discriminative training algorithms. In some cases, the solution can be computed in closed form as in naive Bayes and linear discriminant analysis.

Generalizations of Supervised Learning

There are several ways in which the standard supervised learning problem can be generalized:

1. Semi-supervised learning: In this setting, the desired output values are provided only for a subset of the training data. The remaining data is unlabeled.

2. Active learning: Instead of assuming that all of the training examples are given at the start, active learning algorithms interactively collect new examples, typically by making queries to a human user. Often, the queries are based on unlabeled data, which is a scenario that combines semi-supervised learning with active learning.

3. Structured prediction: When the desired output value is a complex object, such as a parse tree or a labeled graph, then standard methods must be extended.

4. Learning to rank: When the input is a set of objects and the desired output is a ranking of those objects, then again the standard methods must be extended.

Approaches and Algorithms

- Analytical learning
- Artificial neural network
- Backpropagation
- Boosting (meta-algorithm)
- Bayesian statistics
- Case-based reasoning
- Decision tree learning
- Inductive logic programming
- Gaussian process regression
- Group method of data handling
- Kernel estimators
- Learning Automata
- Learning Classifier Systems
- Minimum message length (decision trees, decision graphs, etc.)
- Multilinear subspace learning
- Naive bayes classifier
- Maximum entropy classifier
- Conditional random field
- Nearest Neighbor Algorithm

- Probably approximately correct learning (PAC) learning
- Ripple down rules, a knowledge acquisition methodology
- Symbolic machine learning algorithms
- Subsymbolic machine learning algorithms
- Support vector machines
- Minimum Complexity Machines (MCM)
- Random Forests
- Ensembles of Classifiers
- Ordinal classification
- Data Pre-processing
- Handling imbalanced datasets
- Statistical relational learning
- Proaftn, a multicriteria classification algorithm

Applications

- Bioinformatics
- Cheminformatics
  - Quantitative structure–activity relationship
- Database marketing
- Handwriting recognition
- Information retrieval
  - Learning to rank
- Information extraction
- Object recognition in computer vision
- Optical character recognition
- Spam detection
- Pattern recognition
- Speech recognition
- Supervised learning is a special case of Downward causation in biological systems

General Issues

- Computational learning theory
- Inductive bias
- Overfitting (machine learning)
- (Uncalibrated) Class membership probabilities
- Unsupervised learning
- Version spaces

Semi-supervised Learning

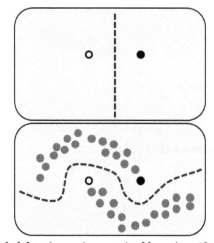

An example of the influence of unlabeled data in semi-supervised learning. The top panel shows a decision boundary we might adopt after seeing only one positive (white circle) and one negative (black circle) example. The bottom panel shows a decision boundary we might adopt if, in addition to the two labeled examples, we were given a collection of unlabeled data (gray circles). This could be viewed as performing clustering and then labeling the clusters with the labeled data, pushing the decision boundary away from high-density regions, or learning an underlying one-dimensional manifold where the data reside.

Semi-supervised learning is a class of supervised learning tasks and techniques that also make use of unlabeled data for training – typically a small amount of labeled data with a large amount of unlabeled data. Semi-supervised learning falls between unsupervised learning (without any labeled training data) and supervised learning (with completely labeled training data). Many machine-learning researchers have found that unlabeled data, when used in conjunction with a small amount of labeled data, can produce considerable improvement in learning accuracy. The acquisition of labeled data for a learning problem often requires a skilled human agent (e.g. to transcribe an audio segment) or a physical experiment (e.g. determining the 3D structure of a protein or determining whether there is oil at a particular location). The cost associated with the labeling process thus may render a fully labeled training set infeasible, whereas acquisition of unlabeled data is relatively inexpensive. In such situations, semi-supervised learning can be of great practical value. Semi-supervised learning is also of theoretical interest in machine learning and as a model for human learning.

As in the supervised learning framework, we are given a set of l independently identically distributed examples $x_1, \ldots, x_l \in X$ with corresponding labels $y_1, \ldots, y_l \in Y$. Additionally, we are given u unlabeled examples $x_{l+1}, \ldots, x_{l+u} \in X$. Semi-supervised learning attempts to make use of this combined information to surpass the classification performance that could be obtained either by discarding the unlabeled data and doing supervised learning or by discarding the labels and doing unsupervised learning.

Semi-supervised learning may refer to either transductive learning or inductive learning. The goal of transductive learning is to infer the correct labels for the given unlabeled data $x_{l+1}, \ldots, x_{l+u}$ only. The goal of inductive learning is to infer the correct mapping from X to Y.

Intuitively, we can think of the learning problem as an exam and labeled data as the few example problems that the teacher solved in class. The teacher also provides a set of unsolved problems. In the transductive setting, these unsolved problems are a take-home exam and you want to do well on them in particular. In the inductive setting, these are practice problems of the sort you will encounter on the in-class exam.

It is unnecessary (and, according to Vapnik's principle, imprudent) to perform transductive learning by way of inferring a classification rule over the entire input space; however, in practice, algorithms formally designed for transduction or induction are often used interchangeably.

Assumptions Used

In order to make any use of unlabeled data, we must assume some structure to the underlying distribution of data. Semi-supervised learning algorithms make use of at least one of the following assumptions.

Smoothness Assumption

Points which are close to each other are more likely to share a label. (More accurately, this is a continuity assumption rather than a smoothness assumption.) This is also generally assumed in supervised learning and yields a preference for geometrically simple decision boundaries. In the case of semi-supervised learning, the smoothness assumption additionally yields a preference for decision boundaries in low-density regions, so that there are fewer points close to each other but in different classes.

Cluster Assumption

The data tend to form discrete clusters, and points in the same cluster are more likely to share a label (although data sharing a label may be spread across multiple clusters). This is a special case of the smoothness assumption and gives rise to feature learning with clustering algorithms.

Manifold Assumption

The data lie approximately on a manifold of much lower dimension than the input space. In this case we can attempt to learn the manifold using both the labeled and unlabeled data to avoid the curse of dimensionality. Then learning can proceed using distances and densities defined on the manifold.

The manifold assumption is practical when high-dimensional data are being generated by some process that may be hard to model directly, but which only has a few degrees of freedom. For instance, human voice is controlled by a few vocal folds, and images of various facial expressions are controlled by a few muscles. We would like in these cases to use distances and smoothness in the natural space of the generating problem, rather than in the space of all possible acoustic waves or images respectively.

History

The heuristic approach of *self-training* (also known as *self-learning* or *self-labeling*) is historically the oldest approach to semi-supervised learning, with examples of applications starting in the 1960s.

The transductive learning framework was formally introduced by Vladimir Vapnik in the 1970s. Interest in inductive learning using generative models also began in the 1970s. A *probably approximately correct* learning bound for semi-supervised learning of a Gaussian mixture was demonstrated by Ratsaby and Venkatesh in 1995.

Semi-supervised learning has recently become more popular and practically relevant due to the variety of problems for which vast quantities of unlabeled data are available—e.g. text on websites, protein sequences, or images. For a review of recent work.

Methods

Generative Models

Generative approaches to statistical learning first seek to estimate $p(x \mid y)$, the distribution of data points belonging to each class. The probability $p(y \mid x)$ that a given point x has label y is then proportional to $p(x \mid y)p(y)$ by Bayes' rule. Semi-supervised learning with generative models can be viewed either as an extension of supervised learning (classification plus information about $p(x)$) or as an extension of unsupervised learning (clustering plus some labels).

Generative models assume that the distributions take some particular form $p(x \mid y, \theta)$ parameterized by the vector θ. If these assumptions are incorrect, the unlabeled data may actually decrease the accuracy of the solution relative to what would have been obtained from labeled data alone. However, if the assumptions are correct, then the unlabeled data necessarily improves performance.

The unlabeled data are distributed according to a mixture of individual-class distributions. In order to learn the mixture distribution from the unlabeled data, it must be identifiable, that is, different parameters must yield different summed distributions. Gaussian mixture distributions are identifiable and commonly used for generative models.

The parameterized joint distribution can be written as $p(x, y \mid \theta) = p(y \mid \theta)p(x \mid y, \theta)$ by using the Chain rule. Each parameter vector θ is associated with a decision function $f_\theta(x) = \underset{y}{\mathrm{argmax}}\, p(y \mid x, \theta)$.

The parameter is then chosen based on fit to both the labeled and unlabeled data, weighted by λ:

$$\underset{\Theta}{\mathrm{argmax}} \left(\log p(\{x_i, y_i\}_{i=1}^{l} \mid \theta) + \lambda \log p(\{x_i\}_{i=l+1}^{l+u} \mid \theta) \right)$$

Low-density Separation

Another major class of methods attempts to place boundaries in regions where there are few data points (labeled or unlabeled). One of the most commonly used algorithms is the transductive support vector machine, or TSVM (which, despite its name, may be used for inductive learning as well). Whereas support vector machines for supervised learning seek a decision boundary with maximal margin over the labeled data, the goal of TSVM is a labeling of the unlabeled data such that the decision boundary has maximal margin over all of the data. In addition to the standard hinge loss $(1 - yf(x))_+$ for labeled data, a loss function $(1 - |f(x)|)_+$ is introduced over the unlabeled data by letting $y = \text{sign} f(x)$. TSVM then selects $f^*(x) = h^*(x) + b$ from a reproducing kernel Hilbert space $\mathcal{H}$ by minimizing the regularized empirical risk:

$$f^* = \underset{f}{\text{argmin}} \left(\sum_{i=1}^{l} (1 - y_i f(x_i))_+ + \lambda_1 \| h \|_{\mathcal{H}}^2 + \lambda_2 \sum_{i=l+1}^{l+u} (1 - |f(x_i)|)_+ \right)$$

An exact solution is intractable due to the non-convex term $(1 - |f(x)|)_+$, so research has focused on finding useful approximations.

Other approaches that implement low-density separation include Gaussian process models, information regularization, and entropy minimization (of which TSVM is a special case).

Graph-based Methods

Graph-based methods for semi-supervised learning use a graph representation of the data, with a node for each labeled and unlabeled example. The graph may be constructed using domain knowledge or similarity of examples; two common methods are to connect each data point to its k nearest neighbors or to examples within some distance ϵ. The weight W_{ij} of an edge between x_i and x_j is then set to $e^{\frac{-\|x_i - x_j\|^2}{\epsilon}}$.

Within the framework of manifold regularization, the graph serves as a proxy for the manifold. A term is added to the standard Tikhonov regularization problem to enforce smoothness of the solution relative to the manifold (in the intrinsic space of the problem) as well as relative to the ambient input space. The minimization problem becomes

$$\underset{f \in \mathcal{H}}{\text{argmin}} \left(\frac{1}{l} \sum_{i=1}^{l} V(f(x_i), y_i) + \lambda_A \| f \|_{\mathcal{H}}^2 + \lambda_I \int_{\mathcal{M}} \| \nabla_{\mathcal{M}} f(x) \|^2 dp(x) \right)$$

where $\mathcal{H}$ is a reproducing kernel Hilbert space and   is the manifold on which the data lie. The regularization parameters λ_A and λ_I control smoothness in the ambient and intrinsic spaces respectively. The graph is used to approximate the intrinsic regularization term. Defining the graph Laplacian $L = D - W$ where $D_{ii} = \sum_{j=1}^{l+u} W_{ij}$ and $\mathbf{f}$ the vector $[f(x_1) \ldots f(x_{l+u})]$, we have

$$\mathbf{f}^T L \mathbf{f} = \sum_{i,j=1}^{l+u} W_{ij} (f_i - f_j)^2 \approx \int_{\mathcal{M}} \| \nabla_{\mathcal{M}} f(x) \|^2 dp(x).$$

The Laplacian can also be used to extend the supervised learning algorithms : regularized least squares and support vector machines (SVM) to semi-supervised versions Laplacian regularized least squares and Laplacian SVM.

Heuristic Approaches

Some methods for semi-supervised learning are not intrinsically geared to learning from both unlabeled and labeled data, but instead make use of unlabeled data within a supervised learning framework. For instance, the labeled and unlabeled examples $x_1, \ldots, x_{l+u}$ may inform a choice of representation, distance metric, or kernel for the data in an unsupervised first step. Then supervised learning proceeds from only the labeled examples.

Self-training is a wrapper method for semi-supervised learning. First a supervised learning algorithm is trained based on the labeled data only. This classifier is then applied to the unlabeled data to generate more labeled examples as input for the supervised learning algorithm. Generally only the labels the classifier is most confident of are added at each step.

Co-training is an extension of self-training in which multiple classifiers are trained on different (ideally disjoint) sets of features and generate labeled examples for one another.

In Human Cognition

Human responses to formal semi-supervised learning problems have yielded varying conclusions about the degree of influence of the unlabeled data. More natural learning problems may also be viewed as instances of semi-supervised learning. Much of human concept learning involves a small amount of direct instruction (e.g. parental labeling of objects during childhood) combined with large amounts of unlabeled experience (e.g. observation of objects without naming or counting them, or at least without feedback).

Human infants are sensitive to the structure of unlabeled natural categories such as images of dogs and cats or male and female faces. More recent work has shown that infants and children take into account not only the unlabeled examples available, but the sampling process from which labeled examples arise.

Active Learning (Machine Learning)

Active learning is a special case of semi-supervised machine learning in which a learning algorithm is able to interactively query the user (or some other information source) to obtain the desired outputs at new data points. In statistics literature it is sometimes also called optimal experimental design.

There are situations in which unlabeled data is abundant but manually labeling is expensive. In such a scenario, learning algorithms can actively query the user/teacher for labels. This type of iterative supervised learning is called active learning. Since the learner chooses the examples, the number of examples to learn a concept can often be much lower than the number required in normal supervised learning. With this approach, there is a risk that the algorithm be overwhelmed by uninformative examples. Recent developments are dedicated to hybrid active learning and active learning in a single-pass (on-line) context, combining concepts from the field of Machine Learning (e.g., conflict and ignorance) with adaptive, incremental learning policies in the field of Online machine learning.

Definitions

Let T be the total set of all data under consideration. For example, in a protein engineering problem, T would include all proteins that are known to have a certain interesting activity and all additional proteins that one might want to test for that activity.

During each iteration, i, T is broken up into three subsets

1. $T_{K,i}$: Data points where the label is known.

2. $T_{U,i}$: Data points where the label is unknown.

3. $T_{C,i}$: A subset of $T_{U,i}$ that is chosen to be labeled.

Most of the current research in active learning involves the best method to choose the data points for $T_{C,i}$.

Query Strategies

Algorithms for determining which data points should be labeled can be organized into a number of different categories:

- Uncertainty sampling: label those points for which the current model is least certain as to what the correct output should be

- Query by committee: a variety of models are trained on the current labeled data, and vote on the output for unlabeled data; label those points for which the "committee" disagrees the most

- Expected model change: label those points that would most change the current model

- Expected error reduction: label those points that would most reduce the model's generalization error

- Variance reduction: label those points that would minimize output variance, which is one of the components of error

- Balance exploration and exploitation: the choice of examples to label is seen as a dilemma between the exploration and the exploitation over the data space representation. This strategy manages this compromise by modelling the active learning problem as a contextual bandit problem. For example, Bouneffouf et al. propose a sequential algorithm named Active Thompson Sampling (ATS), which, in each round, assigns a sampling distribution on the pool, samples one point from this distribution, and queries the oracle for this sample point label.

- Exponentiated Gradient Exploration for Active Learning: In this paper, the author proposes a sequential algorithm named exponentiated gradient (EG)-active that can improve any active learning algorithm by an optimal random exploration.

A wide variety of algorithms have been studied that fall into these categories.

Minimum Marginal Hyperplane

Some active learning algorithms are built upon Support vector machines (SVMs) and exploit the structure of the SVM to determine which data points to label. Such methods usually calculate the margin, W, of each unlabeled datum in $T_{U,i}$ and treat W as an $n-$ dimensional distance from that datum to the separating hyperplane.

Minimum Marginal Hyperplane methods assume that the data with the smallest W are those that the SVM is most uncertain about and therefore should be placed in $T_{C,i}$ to be labeled. Other similar methods, such as Maximum Marginal Hyperplane, choose data with the largest W. Tradeoff methods choose a mix of the smallest and largest Ws.

Structured Prediction

Structured prediction or structured (output) learning is an umbrella term for supervised machine learning techniques that involves predicting structured objects, rather than scalar discrete or real values.

For example, the problem of translating a natural language sentence into a syntactic representation such as a parse tree can be seen as a structured prediction problem in which the structured output domain is the set of all possible parse trees.

Probabilistic graphical models form a large class of structured prediction models. In particular, Bayesian networks and random fields are popularly used to solve structured prediction problems in a wide variety of application domains including bioinformatics, natural language processing, speech recognition, and computer vision. Other algorithms and models for structured prediction include inductive logic programming, case-based reasoning, structured SVMs, Markov logic networks and constrained conditional models.

Similar to commonly used supervised learning techniques, structured prediction models are typically trained by means of observed data in which the true prediction value is used to adjust model parameters. Due to the complexity of the model and the interrelations of predicted variables the process of prediction using a trained model and of training itself is often computationally infeasible and approximate inference and learning methods are used.

Example: Sequence Tagging

Sequence tagging is a class of problems prevalent in natural language processing, where input data are often sequences (e.g. sentences of text). The sequence tagging problem appears in several guises, e.g. part-of-speech tagging and named entity recognition. In POS tagging, for example, each word in a sequence must receive a "tag" (class label) that expresses its "type" of word:

- This DT

- is VBZ

- a DT

- tagged JJ

- sentence NN

The main challenge of this problem is to resolve ambiguity: the word "sentence" can also be a verb in English, and so can "tagged".

While this problem can be solved by simply performing classification of individual tokens, that approach does not take into account the empirical fact that tags do not occur independently; instead, each tag displays a strong conditional dependence on the tag of the previous word. This fact can be exploited in a sequence model such as a hidden Markov model or conditional random field that predicts the entire tag sequence for a sentence, rather than just individual tags, by means of the Viterbi algorithm.

Structured Perceptron

One of the easiest ways to understand algorithms for general structured prediction is the structured perceptron of Collins. This algorithm combines the perceptron algorithm for learning linear classifiers with an inference algorithm (classically the Viterbi algorithm when used on sequence data) and can be described abstractly as follows. First define a "joint feature function" $\Phi(x, y)$ that maps a training sample x and a candidate prediction y to a vector of length n (x and y may have any structure; n is problem-dependent, but must be fixed for each model). Let GEN be a function that generates candidate predictions. Then:

Let w be a weight vector of length n

For a pre-determined number of iterations:

For each sample x in the training set with true output t:

Make a prediction $\hat{y} = \arg\max\{y \in GEN(x)\}(w^T \phi(x, y))$

Update w, from $\hat{y}$ to t: $w = w + c(-\phi(x, \hat{y}) + \phi(x, t))$, c is learning rate

In practice, finding the argmax over $GEN(x)$ will be done using an algorithm such as Viterbi or an algorithm such as max-sum, rather than an exhaustive search through an exponentially large set of candidates.

The idea of learning is similar to multiclass perceptron.

Learning to Rank

Learning to rank or machine-learned ranking (MLR) is the application of machine learning, typically supervised, semi-supervised or reinforcement learning, in the construction of ranking models for information retrieval systems. Training data consists of lists of items with some partial order specified between items in each list. This order is typically induced by giving a numerical or ordinal score or a binary judgment (e.g. "relevant" or "not relevant") for each item. The ranking model's purpose is to rank, i.e. produce a permutation of items in new, unseen lists in a way which is "similar" to rankings in the training data in some sense.

Applications

In Information Retrieval

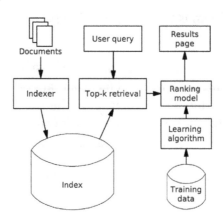

A possible architecture of a machine-learned search engine.

Ranking is a central part of many information retrieval problems, such as document retrieval, collaborative filtering, sentiment analysis, and online advertising.

A possible architecture of a machine-learned search engine is shown in the figure.

Training data consists of queries and documents matching them together with relevance degree of each match. It may be prepared manually by human *assessors* (or *raters*, as Google calls them), who check results for some queries and determine relevance of each result. It is not feasible to check relevance of all documents, and so typically a technique called pooling is used — only the top few documents, retrieved by some existing ranking models are checked. Alternatively, training data may be derived automatically by analyzing *clickthrough logs* (i.e. search results which got clicks from users), *query chains*, or such search engines' features as Google's SearchWiki.

Training data is used by a learning algorithm to produce a ranking model which computes relevance of documents for actual queries.

Typically, users expect a search query to complete in a short time (such as a few hundred milliseconds for web search), which makes it impossible to evaluate a complex ranking model on each document in the corpus, and so a two-phase scheme is used. First, a small number of potentially relevant documents are identified using simpler retrieval models which permit fast query evaluation, such as the vector space model, boolean model, weighted AND, or BM25. This phase is called *top-k document retrieval* and many heuristics were proposed in the literature to accelerate it, such as using a document's static quality score and tiered indexes. In the second phase, a more accurate but computationally expensive machine-learned model is used to re-rank these documents.

In other Areas

Learning to rank algorithms have been applied in areas other than information retrieval:

- In machine translation for ranking a set of hypothesized translations;

- In computational biology for ranking candidate 3-D structures in protein structure prediction problem.

- In Recommender systems for identifying a ranked list of related news articles to recommend to a user after he or she has read a current news article.

Feature Vectors

For convenience of MLR algorithms, query-document pairs are usually represented by numerical vectors, which are called *feature vectors*. Such an approach is sometimes called *bag of features* and is analogous to the bag of words model and vector space model used in information retrieval for representation of documents.

Components of such vectors are called *features*, *factors* or *ranking signals*. They may be divided into three groups (features from document retrieval are shown as examples):

- *Query-independent* or *static* features — those features, which depend only on the document, but not on the query. For example, PageRank or document's length. Such features can be precomputed in off-line mode during indexing. They may be used to compute document's *static quality score* (or *static rank*), which is often used to speed up search query evaluation.

- *Query-dependent* or *dynamic* features — those features, which depend both on the contents of the document and the query, such as TF-IDF score or other non-machine-learned ranking functions.

- *Query level features* or *query features*, which depend only on the query. For example, the number of words in a query.

Some examples of features, which were used in the well-known LETOR dataset:

- TF, TF-IDF, BM25, and language modeling scores of document's zones (title, body, anchors text, URL) for a given query;

- Lengths and IDF sums of document's zones;

- Document's PageRank, HITS ranks and their variants.

Selecting and designing good features is an important area in machine learning, which is called feature engineering.

Evaluation Measures

There are several measures (metrics) which are commonly used to judge how well an algorithm is doing on training data and to compare performance of different MLR algorithms. Often a learning-to-rank problem is reformulated as an optimization problem with respect to one of these metrics.

Examples of ranking quality measures:

- Mean average precision (MAP);

- DCG and NDCG;

- Precision@n, NDCG@n, where "@n" denotes that the metrics are evaluated only on top n documents;

- Mean reciprocal rank;

- Kendall's tau

- Spearman's Rho

DCG and its normalized variant NDCG are usually preferred in academic research when multiple levels of relevance are used. Other metrics such as MAP, MRR and precision, are defined only for binary judgements.

Recently, there have been proposed several new evaluation metrics which claim to model user's satisfaction with search results better than the DCG metric:

- Expected reciprocal rank (ERR);

- Yandex's pfound.

Both of these metrics are based on the assumption that the user is more likely to stop looking at search results after examining a more relevant document, than after a less relevant document.

Approaches

Tie-Yan Liu of Microsoft Research Asia has analyzed existing algorithms for learning to rank problems in his paper "Learning to Rank for Information Retrieval". He categorized them into three groups by their input representation and loss function:

Pointwise Approach

In this case it is assumed that each query-document pair in the training data has a numerical or ordinal score. Then learning-to-rank problem can be approximated by a regression problem — given a single query-document pair, predict its score.

A number of existing supervised machine learning algorithms can be readily used for this purpose. Ordinal regression and classification algorithms can also be used in pointwise approach when they are used to predict score of a single query-document pair, and it takes a small, finite number of values.

Pairwise Approach

In this case learning-to-rank problem is approximated by a classification problem — learning a binary classifier that can tell which document is better in a given pair of documents. The goal is to minimize average number of inversions in ranking.

Listwise Approach

These algorithms try to directly optimize the value of one of the above evaluation measures, averaged over all queries in the training data. This is difficult because most evaluation measures are not continuous functions with respect to ranking model's parameters, and so continuous approximations or bounds on evaluation measures have to be used.

History

Norbert Fuhr introduced the general idea of MLR in 1992, describing learning approaches in information retrieval as a generalization of parameter estimation; a specific variant of this approach (using polynomial regression) had been published by him three years earlier. Bill Cooper proposed logistic regression for the same purpose in 1992 and used it with his Berkeley research group to train a successful ranking function for TREC. Manning et al. suggest that these early works achieved limited results in their time due to little available training data and poor machine learning techniques.

Several conferences, such as NIPS, SIGIR and ICML had workshops devoted to the learning-to-rank problem since mid-2000s (decade).

Practical Usage by Search Engines

Commercial web search engines began using machine learned ranking systems since the 2000s (decade). One of the first search engines to start using it was AltaVista (later its technology was acquired by Overture, and then Yahoo), which launched a gradient boosting-trained ranking function in April 2003.

Bing's search is said to be powered by RankNet algorithm, which was invented at Microsoft Research in 2005.

In November 2009 a Russian search engine Yandex announced that it had significantly increased its search quality due to deployment of a new proprietary MatrixNet algorithm, a variant of gradient boosting method which uses oblivious decision trees. Recently they have also sponsored a machine-learned ranking competition "Internet Mathematics 2009" based on their own search engine's production data. Yahoo has announced a similar competition in 2010.

As of 2008, Google's Peter Norvig denied that their search engine exclusively relies on machine-learned ranking. Cuil's CEO, Tom Costello, suggests that they prefer hand-built models because they can outperform machine-learned models when measured against metrics like click-through rate or time on landing page, which is because machine-learned models "learn what people say they like, not what people actually like".

In January 2017 the technology was included in the open source search engine Apache Solr™, thus making machine learned search rank widely accessible also for enterprise search.

Bias–variance Tradeoff

A function (red) is approximated using radial basis functions (blue). Several trials are shown in each graph. For each trial, a few noisy data points are provided as training set (top). For a wide spread (image 2) the bias is high: the RBFs cannot fully approximate the function (especially the central dip), but the variance between different trials is low. As spread decreases (image 3 and 4) the bias decreases: the blue curves more closely approximate the red. However, depending on the noise in different trials the variance between trials increases. In the lowermost image the approximated values for x=0 varies wildly depending on where the data points were located.

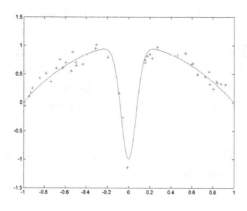

Function and noisy data.

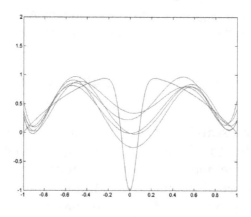

Spread=5

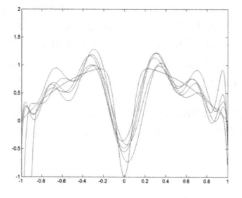

Spread=1

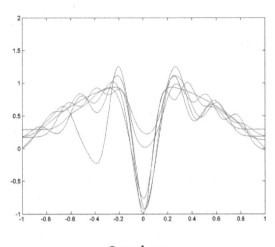

Spread=0.1

In statistics and machine learning, the bias–variance tradeoff (or dilemma) is the problem of simultaneously minimizing two sources of error that prevent supervised learning algorithms from generalizing beyond their training set:

- The *bias* is error from erroneous assumptions in the learning algorithm. High bias can cause an algorithm to miss the relevant relations between features and target outputs (underfitting).

- The *variance* is error from sensitivity to small fluctuations in the training set. High variance can cause overfitting: modeling the random noise in the training data, rather than the intended outputs.

The bias–variance decomposition is a way of analyzing a learning algorithm's expected generalization error with respect to a particular problem as a sum of three terms, the bias, variance, and a quantity called the *irreducible error*, resulting from noise in the problem itself.

This tradeoff applies to all forms of supervised learning: classification, regression (function fitting), and structured output learning. It has also been invoked to explain the effectiveness of heuristics in human learning.

Motivation

The bias–variance tradeoff is a central problem in supervised learning. Ideally, one wants to choose a model that both accurately captures the regularities in its training data, but also generalizes well to unseen data. Unfortunately, it is typically impossible to do both simultaneously. High-variance learning methods may be able to represent their training set well, but are at risk of overfitting to noisy or unrepresentative training data. In contrast, algorithms with high bias typically produce simpler models that don't tend to overfit, but may *underfit* their training data, failing to capture important regularities.

Models with low bias are usually more complex (e.g. higher-order regression polynomials), enabling them to represent the training set more accurately. In the process, however, they may also represent a large noise component in the training set, making their predictions less accurate -

despite their added complexity. In contrast, models with higher bias tend to be relatively simple (low-order or even linear regression polynomials), but may produce lower variance predictions when applied beyond the training set.

Bias–variance Decomposition of Squared Error

Suppose that we have a training set consisting of a set of points $x_1,\ldots,x_n$ and real values y_i associated with each point x_i. We assume that there is a function with noise $y = f(x) + \epsilon$, where the noise, ϵ, has zero mean and variance σ^2.

We want to find a function $\hat{f}(x)$, that approximates the true function $f(x)$ as well as possible, by means of some learning algorithm. We make "as well as possible" precise by measuring the mean squared error between y and $\hat{f}(x)$: we want $(y - \hat{f}(x))^2$ to be minimal, both for $x_1,\ldots,x_n$ *and for points outside of our sample*. Of course, we cannot hope to do so perfectly, since the y_i contain noise ϵ; this means we must be prepared to accept an *irreducible error* in any function we come up with.

Finding an $\hat{f}$ that generalizes to points outside of the training set can be done with any of the countless algorithms used for supervised learning. It turns out that whichever function $\hat{f}$ we select, we can decompose its expected error on an unseen sample x as follows:

$$E\left[(y - \hat{f}(x))^2\right] = \text{Bias}\left[\hat{f}(x)\right]^2 + \text{Var}\left[\hat{f}(x)\right] + \sigma^2$$

Where:

$$\text{Bias}\left[\hat{f}(x)\right] = E\left[\hat{f}(x) - f(x)\right]$$

and

$$\text{Var}\left[\hat{f}(x)\right] = E[\hat{f}(x)^2] - E[\hat{f}(x)]^2$$

The expectation ranges over different choices of the training set $x_1,\ldots,x_n, y_1,\ldots,y_n$, all sampled from the same joint distribution $P(x, y)$. The three terms represent:

- the square of the *bias* of the learning method, which can be thought of as the error caused by the simplifying assumptions built into the method. E.g., when approximating a non-linear function $f(x)$ using a learning method for linear models, there will be error in the estimates $\hat{f}(x)$ due to this assumption;

- the *variance* of the learning method, or, intuitively, how much the learning method $\hat{f}(x)$ will move around its mean;

- the irreducible error σ^2. Since all three terms are non-negative, this forms a lower bound on the expected error on unseen samples.

The more complex the model $\hat{f}(x)$ is, the more data points it will capture, and the lower the bias will be. However, complexity will make the model "move" more to capture the data points, and hence its variance will be larger.

Derivation

The derivation of the bias–variance decomposition for squared error proceeds as follows. For notational convenience, abbreviate $f = f(x)$ and $\hat{f} = \hat{f}(x)$. First, note that for any random variable X, we have

$$Var[X] = E[X^2] - E[X]^2$$

Rearranging, we get:

$$E[X^2] = Var[X] + E[X]^2$$

Since f is deterministic

$$E[f] \quad f.$$

This, given $y = f + \epsilon$ and $E[\epsilon] = 0$, implies $E[y] = E[f + \epsilon] = E[f] = f$.

Also, since $Var[\epsilon] = \sigma^2$

$$Var[y] = E[(y - E[y])^2] = E[(y - f)^2] = E[(f + \epsilon - f)^2] = E[\epsilon^2] = Var[\epsilon] + E[\epsilon]^2 = \sigma^2$$

Thus, since ϵ and $\hat{f}$ are independent, we can write

$$\begin{aligned}
E\left[(y - \hat{f})^2\right] &= E[y^2 + \hat{f}^2 - 2y\hat{f}] \\
&= E[y^2] + E[\hat{f}^2] - E[2y\hat{f}] \\
&= Var[y] + E[y]^2 + Var[\hat{f}] + E[\hat{f}]^2 - 2fE[\hat{f}] \\
&= Var[y] + Var[\hat{f}] + (f^2 - 2fE[\hat{f}] + E[\hat{f}]^2) \\
&= Var[y] + Var[\hat{f}] + (f - E[\hat{f}])^2 \\
&= Var[y] + Var[\hat{f}] + E[f - \hat{f}]^2 \\
&= \sigma^2 + Var[\hat{f}] + Bias[\hat{f}]^2
\end{aligned}$$

Application to Regression

The bias-variance decomposition forms the conceptual basis for regression regularization methods such as Lasso and Ridge Regression. Regularization methods introduce bias into the regression solution that can reduce variance considerably relative to the OLS solution. Although the OLS solution provides non-biased regression estimates, the lower variance solutions produced by regularization techniques provide superior MSE performance.

Application to Classification

The bias–variance decomposition was originally formulated for least-squares regression. For the case of classification under the 0-1 loss (misclassification rate), it's possible to find a similar decomposition. Alternatively, if the classification problem can be phrased as probabilistic classification, then the expected squared error of the predicted probabilities with respect to the true probabilities can be decomposed as before.

Approaches

Dimensionality reduction and feature selection can decrease variance by simplifying models. Similarly, a larger training set tends to decrease variance. Adding features (predictors) tends to decrease bias, at the expense of introducing additional variance. Learning algorithms typically have some tunable parameters that control bias and variance, e.g.:

- (Generalized) linear models can be regularized to decrease their variance at the cost of increasing their bias.

- In artificial neural networks, the variance increases and the bias decreases with the number of hidden units. Like in GLMs, regularization is typically applied.

- In k-nearest neighbor models, a high value of k leads to high bias and low variance.

- In Instance-based learning, regularization can be achieved varying the mixture of prototypes and exemplars.

- In decision trees, the depth of the tree determines the variance. Decision trees are commonly pruned to control variance.

One way of resolving the trade-off is to use mixture models and ensemble learning. For example, boosting combines many "weak" (high bias) models in an ensemble that has lower bias than the individual models, while bagging combines "strong" learners in a way that reduces their variance.

K-nearest Neighbors

In the case of k-nearest neighbors regression, a closed-form expression exists that relates the bias–variance decomposition to the parameter k:

$$E[(y - \hat{f}(x))^2] = \left(f(x) - \frac{1}{k} \sum_{i=1}^{k} f(N_i(x)) \right)^2 + \frac{\sigma^2}{k} + \sigma^2$$

where $N_1(x), \ldots, N_k(x)$ are the k nearest neighbors of x in the training set. The bias (first term) is a monotone rising function of k, while the variance (second term) drops off as k is increased. In fact, under "reasonable assumptions" the bias of the first-nearest neighbor (1-NN) estimator vanishes entirely as the size of the training set approaches infinity.

Application to Human Learning

While widely discussed in the context of machine learning, the bias-variance dilemma has been examined in the context of human cognition, most notably by Gerd Gigerenzer and co-workers in the context of learned heuristics. They have argued that the human brain resolves the dilemma in the case of the typically sparse, poorly-characterised training-sets provided by experience by adopting high-bias/low variance heuristics. This reflects the fact that a zero-bias approach has poor generalisability to new situations, and also unreasonably presumes precise knowledge of the true state of the world. The resulting heuristics are relatively simple, but produce better inferences in a wider variety of situations.

Geman et al. argue that the bias-variance dilemma implies that abilities such as generic object recognition cannot be learned from scratch, but require a certain degree of "hard wiring" that is later tuned by experience. This is because model-free approaches to inference require impractically large training sets if they are to avoid high variance.

Perceptron

In machine learning, the perceptron is an algorithm for supervised learning of binary classifiers (functions that can decide whether an input, represented by a vector of numbers, belongs to some specific class or not). It is a type of linear classifier, i.e. a classification algorithm that makes its predictions based on a linear predictor function combining a set of weights with the feature vector. The algorithm allows for online learning, in that it processes elements in the training set one at a time.

The perceptron algorithm dates back to the late 1950s; its first implementation, in custom hardware, was one of the first artificial neural networks to be produced.

History

The Mark I Perceptron machine was the first implementation of the perceptron algorithm. The machine was connected to a camera that used 20×20 cadmium sulfide photocells to produce a 400-pixel image. The main visible feature is a patchboard that allowed experimentation with different combinations of input features. To the right of that are arrays of potentiometers that implemented the adaptive weights.

The perceptron algorithm was invented in 1957 at the Cornell Aeronautical Laboratory by Frank Rosenblatt, funded by the United States Office of Naval Research. The perceptron was intended to be a machine, rather than a program, and while its first implementation was in software for the IBM 704, it was subsequently implemented in custom-built hardware as the "Mark 1 perceptron". This machine was designed for image recognition: it had an array of 400 photocells, randomly connected to the "neurons". Weights were encoded in potentiometers, and weight updates during learning were performed by electric motors.

In a 1958 press conference organized by the US Navy, Rosenblatt made statements about the perceptron that caused a heated controversy among the fledgling AI community; based on Rosenblatt's statements, *The New York Times* reported the perceptron to be "the embryo of an electronic computer that [the Navy] expects will be able to walk, talk, see, write, reproduce itself and be conscious of its existence."

Although the perceptron initially seemed promising, it was quickly proved that perceptrons could not be trained to recognise many classes of patterns. This caused the field of neural network research to stagnate for many years, before it was recognised that a feedforward neural network with two or more layers (also called a multilayer perceptron) had far greater processing power than perceptrons with one layer (also called a single layer perceptron). Single layer perceptrons are only capable of learning linearly separable patterns; in 1969 a famous book entitled *Perceptrons* by Marvin Minsky and Seymour Papert showed that it was impossible for these classes of network to learn an XOR function. It is often believed that they also conjectured (incorrectly) that a similar result would hold for a multi-layer perceptron network. However, this is not true, as both Minsky and Papert already knew that multi-layer perceptrons were capable of producing an XOR function. Three years later Stephen Grossberg published a series of papers introducing networks capable of modelling differential, contrast-enhancing and XOR functions. (The papers were published in 1972 and 1973, *Grossberg (1973). "Contour enhancement, short-term memory, and constancies in reverberating neural networks". Studies in Applied Mathematics*). Nevertheless, the often-miscited Minsky/Papert text caused a significant decline in interest and funding of neural network research. It took ten more years until neural network research experienced a resurgence in the 1980s. This text was reprinted in 1987 as "Perceptrons - Expanded Edition" where some errors in the original text are shown and corrected.

The kernel perceptron algorithm was already introduced in 1964 by Aizerman et al. Margin bounds guarantees were given for the Perceptron algorithm in the general non-separable case first by Freund and Schapire (1998), and more recently by Mohri and Rostamizadeh (2013) who extend previous results and give new L1 bounds.

Definition

In the modern sense, the perceptron is an algorithm for learning a binary classifier: a function that maps its input x (a real-valued vector) to an output value $f(x)$ (a single binary value):

$$f(x) = \begin{cases} 1 & \text{if } w \cdot x + b > 0 \\ 0 & \text{otherwise} \end{cases}$$

where w is a vector of real-valued weights, $w \cdot x$ is the dot product $\sum_{i=1} w_i x_i$, where m is the number of inputs to the perceptron and b is the *bias*. The bias shifts the decision boundary away from the origin and does not depend on any input value.

The value of $f(x)$ (0 or 1) is used to classify x as either a positive or a negative instance, in the case of a binary classification problem. If b is negative, then the weighted combination of inputs must produce a positive value greater than $|b|$ in order to push the classifier neuron over the 0 threshold. Spatially, the bias alters the position (though not the orientation) of the decision boundary. The perceptron learning algorithm does not terminate if the learning set is not linearly separable. If the vectors are not linearly separable learning will never reach a point where all vectors are classified properly. The most famous example of the perceptron's inability to solve problems with linearly nonseparable vectors is the Boolean exclusive-or problem. The solution spaces of decision boundaries for all binary functions and learning behaviors are studied in the reference.

In the context of neural networks, a perceptron is an artificial neuron using the Heaviside step function as the activation function. The perceptron algorithm is also termed the single-layer perceptron, to distinguish it from a multilayer perceptron, which is a misnomer for a more complicated neural network. As a linear classifier, the single-layer perceptron is the simplest feedforward neural network.

Learning Algorithm

Below is an example of a learning algorithm for a (single-layer) perceptron. For multilayer perceptrons, where a hidden layer exists, more sophisticated algorithms such as backpropagation must be used. Alternatively, methods such as the delta rule can be used if the function is non-linear and differentiable, although the one below will work as well.

When multiple perceptrons are combined in an artificial neural network, each output neuron operates independently of all the others; thus, learning each output can be considered in isolation.

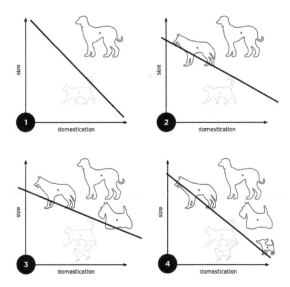

A diagram showing a perceptron updating its linear boundary as more training examples are added.

Definitions

We first define some variables:

- $y = f(\mathbf{z})$ denotes the *output* from the perceptron for an input vector $\mathbf{z}$.

- $D = \{(\mathbf{x}_1, d_1), \ldots, (\mathbf{x}_s, d_s)\}$ is the *training set* of s samples, where:

  o $\mathbf{x}_j$ is the n dimensional input vector.

  o d_j is the desired output value of the perceptron for that input.

We show the values of the features as follows:

- $x_{j,i}$ is the value of the i th feature of the j th training *input vector*.

- $x_{j,0} = 1$.

To represent the weights:

- w_i is the i th value in the *weight vector*, to be multiplied by the value of the i th input feature.

- Because $x_{j,0} = 1$, the w_0 is effectively a bias that we use instead of the bias constant b.

To show the time-dependence of **w**, we use:

- $w_i(t)$ is the weight i at time t.

Unlike other linear classification algorithms such as logistic regression, there is no need for a *learning rate* in the perceptron algorithm. This is because multiplying the update by any constant simply rescales the weights but never changes the sign of the prediction.

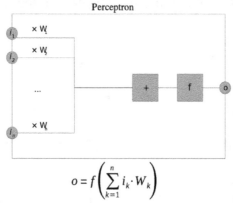

$$o = f\left(\sum_{k=1}^{n} i_k \cdot w_k\right)$$

The appropriate weights are applied to the inputs, and the resulting weighted sum passed to a function that produces the output o.

Steps

1. Initialize the weights and the threshold. Weights may be initialized to 0 or to a small random value. In the example below, we use 0.

2. For each example j in our training set D, perform the following steps over the input x_j and desired output d_j:

    a. Calculate the actual output:

    $$y_j(t) = f[\mathbf{w}(t) \cdot \mathbf{x}_j]$$
    $$= f[w_0(t)x_{j,0} + w_1(t)x_{j,1} + w_2(t)x_{j,2} + \cdots + w_n(t)x_{j,n}]$$

    b. Update the weights:

    $$w_i(t+1) = w_i(t) + (d_j - y_j(t))x_{j,i} \text{ , for all features } 0 \le i \le n.$$

3. For offline learning, the step 2 may be repeated until the iteration error $\dfrac{1}{s}\sum_{j=1}^{s}|d_j - y_j(t)|$ is less than a user-specified error threshold γ, or a predetermined number of iterations have been completed.

The algorithm updates the weights after steps 2a and 2b. These weights are immediately applied to a pair in the training set, and subsequently updated, rather than waiting until all pairs in the training set have undergone these steps.

Convergence

The perceptron is a linear classifier, therefore it will never get to the state with all the input vectors classified correctly if the training set D is not linearly separable, i.e. if the positive examples can not be separated from the negative examples by a hyperplane. In this case, no "approximate" solution will be gradually approached under the standard learning algorithm, but instead learning will fail completely. Hence, if linear separability of the training set is not known a priori, one of the training variants below should be used.

But if the training set *is* linearly separable, then the perceptron is guaranteed to converge, and there is an upper bound on the number of times the perceptron will adjust its weights during the training.

Suppose that the input vectors from the two classes can be separated by a hyperplane with a margin γ, i.e. there exists a weight vector $\mathbf{w}, \|\mathbf{w}\|=1$, and a bias term b such that $\mathbf{w} \cdot \mathbf{x}_j > \gamma$ for all $j : d_j = 1$ and $\mathbf{w} \cdot \mathbf{x}_j < -\gamma$ for all $j : d_j = 0$. And also let R denote the maximum norm of an input vector. Novikoff (1962) proved that in this case the perceptron algorithm converges after making $O(R^2 / \gamma^2)$ updates. The idea of the proof is that the weight vector is always adjusted by a bounded amount in a direction with which it has a negative dot product, and thus can be bounded above by $O(\sqrt{t})$ where t is the number of changes to the weight vector. But it can also be bounded below by $O(t)$ because if there exists an (unknown) satisfactory weight vector, then every change makes progress in this (unknown) direction by a positive amount that depends only on the input vector.

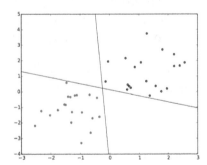

Two classes of points, and two of the infinitely many linear boundaries that separate them. Even though the boundaries are at nearly right angles to one another, the perceptron algorithm has no way of choosing between them.

While the perceptron algorithm is guaranteed to converge on *some* solution in the case of a linearly separable training set, it may still pick *any* solution and problems may admit many solutions of varying quality. The *perceptron of optimal stability*, nowadays better known as the linear support vector machine, was designed to solve this problem.

Variants

The pocket algorithm with ratchet (Gallant, 1990) solves the stability problem of perceptron learning by keeping the best solution seen so far "in its pocket". The pocket algorithm then returns the solution in the pocket, rather than the last solution. It can be used also for non-separable data sets, where the aim is to find a perceptron with a small number of misclassifications. However, these solutions appear purely stochastically and hence the pocket algorithm neither approaches them gradually in the course of learning, nor are they guaranteed to show up within a given number of learning steps.

The Maxover algorithm (Wendemuth, 1995) is "robust" in the sense that it will converge regardless of (prior) knowledge of linear separability of the data set. In the linear separable case, it will solve the training problem – if desired, even with optimal stability (maximum margin between the classes). For non-separable data sets, it will return a solution with a small number of misclassifications. In all cases, the algorithm gradually approaches the solution in the course of learning, without memorizing previous states and without stochastic jumps. Convergence is to global optimality for separable data sets and to local optimality for non-separable data sets.

In separable problems, perceptron training can also aim at finding the largest separating margin between the classes. The so-called perceptron of optimal stability can be determined by means of iterative training and optimization schemes, such as the Min-Over algorithm (Krauth and Mezard, 1987) or the AdaTron (Anlauf and Biehl, 1989)). AdaTron uses the fact that the corresponding quadratic optimization problem is convex. The perceptron of optimal stability, together with the kernel trick, are the conceptual foundations of the support vector machine.

The α-perceptron further used a pre-processing layer of fixed random weights, with thresholded output units. This enabled the perceptron to classify analogue patterns, by projecting them into a binary space. In fact, for a projection space of sufficiently high dimension, patterns can become linearly separable.

Another way to solve nonlinear problems without using multiple layers is to use higher order networks (sigma-pi unit). In this type of network, each element in the input vector is extended with each pairwise combination of multiplied inputs (second order). This can be extended to an n-order network.

It should be kept in mind, however, that the best classifier is not necessarily that which classifies all the training data perfectly. Indeed, if we had the prior constraint that the data come from equi-variant Gaussian distributions, the linear separation in the input space is optimal, and the nonlinear solution is overfitted.

Other linear classification algorithms include Winnow, support vector machine and logistic regression.

Multiclass Perceptron

Like most other techniques for training linear classifiers, the perceptron generalizes naturally to multiclass classification. Here, the input x and the output y are drawn from arbitrary sets. A feature representation function $f(x, y)$ maps each possible input/output pair to a finite-dimensional real-valued feature vector. As before, the feature vector is multiplied by a weight vector w, but now the resulting score is used to choose among many possible outputs:

$$\hat{y} = \text{argmax}_y \, f(x, y) \cdot w .$$

$\approx$ Learning again iterates over the examples, predicting an output for each, leaving the weights unchanged when the predicted output matches the target, and changing them when it does not. The update becomes:

$$w_{t+1} = w_t + f(x, y) - f(x, \hat{y}).$$

This multiclass feedback formulation reduces to the original perceptron when x is a real-valued vector, y is chosen from $\{0,1\}$, and $f(x, y) = yx$.

For certain problems, input/output representations and features can be chosen so that $\text{argmax}_y f(x, y) \cdot w$ can be found efficiently even though y is chosen from a very large or even infinite set.

In recent years, perceptron training has become popular in the field of natural language processing for such tasks as part-of-speech tagging and syntactic parsing (Collins, 2002).

Unsupervised Learning

Unsupervised machine learning is the machine learning task of inferring a function to describe hidden structure from "unlabeled" data (a classification or categorization is not included in the observations). Since the examples given to the learner are unlabeled, there is no evaluation of the accuracy of the structure that is output by the relevant algorithm—which is one way of distinguishing unsupervised learning from supervised learning and reinforcement learning.

A central case of unsupervised learning is the problem of density estimation in statistics, though unsupervised learning encompasses many other problems (and solutions) involving summarizing and explaining key features of the data.

Approaches to unsupervised learning include:

- clustering
  - k-means
  - mixture models
  - hierarchical clustering,
- anomaly detection
- Neural Networks
  - Hebbian Learning
  - Generative Adversarial Networks
- Approaches for learning latent variable models such as
  - Expectation–maximization algorithm (EM)

- o Method of moments

- o Blind signal separation techniques, e.g.,

  - • Principal component analysis,

  - • Independent component analysis,

  - • Non-negative matrix factorization,

  - • Singular value decomposition.

In Neural Networks

The classical example of unsupervised learning in the study of both natural and artificial neural networks is subsumed by Donald Hebb's principle, that is, neurons that fire together wire together. In Hebbian learning, the connection is reinforced irrespective of an error, but is exclusively a function of the coincidence between action potentials between the two neurons. A similar version that modifies synaptic weights takes into account the time between the action potentials (spike-timing-dependent plasticity or STDP). Hebbian Learning has been hypothesized to underlie a range of cognitive functions, such as pattern recognition and experiential learning.

Among neural network models, the self-organizing map (SOM) and adaptive resonance theory (ART) are commonly used unsupervised learning algorithms. The SOM is a topographic organization in which nearby locations in the map represent inputs with similar properties. The ART model allows the number of clusters to vary with problem size and lets the user control the degree of similarity between members of the same clusters by means of a user-defined constant called the vigilance parameter. ART networks are also used for many pattern recognition tasks, such as automatic target recognition and seismic signal processing. The first version of ART was "ART1", developed by Carpenter and Grossberg (1988).

Method of Moments

One of the statistical approaches for unsupervised learning is the method of moments. In the method of moments, the unknown parameters (of interest) in the model are related to the moments of one or more random variables, and thus, these unknown parameters can be estimated given the moments. The moments are usually estimated from samples empirically. The basic moments are first and second order moments. For a random vector, the first order moment is the mean vector, and the second order moment is the covariance matrix (when the mean is zero). Higher order moments are usually represented using tensors which are the generalization of matrices to higher orders as multi-dimensional arrays.

In particular, the method of moments is shown to be effective in learning the parameters of latent variable models. Latent variable models are statistical models where in addition to the observed variables, a set of latent variables also exists which is not observed. A highly practical example of latent variable models in machine learning is the topic modeling which is a statistical model for generating the words (observed variables) in the document based on the topic (latent variable) of the document. In the topic modeling, the words in the document are generated according to different statistical parameters when the topic of the document is changed. It is shown that method of moments (tensor decomposition techniques) consistently recover the parameters of a large class of latent variable models under some assumptions.

The Expectation–maximization algorithm (EM) is also one of the most practical methods for learning latent variable models. However, it can get stuck in local optima, and it is not guaranteed that the algorithm will converge to the true unknown parameters of the model. Alternatively, for the method of moments, the global convergence is guaranteed under some conditions.

Examples

Behavioral-based detection in network security has become a good application area for a combination of supervised- and unsupervised-machine learning. This is because the amount of data for a human security analyst to analyze is impossible (measured in terabytes per day) to review to find patterns and anomalies. According to Giora Engel, co-founder of LightCyber, in a *Dark Reading* article, "The great promise machine learning holds for the security industry is its ability to detect advanced and unknown attacks particularly those leading to data breaches." The basic premise is that a motivated attacker will find their way into a network (generally by compromising a user's computer or network account through phishing, social engineering or malware). The security challenge then becomes finding the attacker by their operational activities, which include reconnaissance, lateral movement, command & control and exfiltration. These activities especially reconnaissance and lateral movement--stand in contrast to an established baseline of "normal" or "good" activity for each user and device on the network. The role of machine learning is to create ongoing profiles for users and devices and then find meaningful anomalies.

Competitive Learning

Competitive learning is a form of unsupervised learning in artificial neural networks, in which nodes compete for the right to respond to a subset of the input data. A variant of Hebbian learning, competitive learning works by increasing the specialization of each node in the network. It is well suited to finding clusters within data.

Models and algorithms based on the principle of competitive learning include vector quantization and self-organizing maps (Kohonen maps).

Principles

There are three basic elements to a competitive learning rule:

- A set of neurons that are all the same except for some randomly distributed synaptic weights, and which therefore respond differently to a given set of input patterns

- A limit imposed on the "strength" of each neuron

- A mechanism that permits the neurons to compete for the right to respond to a given subset of inputs, such that only one output neuron (or only one neuron per group), is active (i.e. "on") at a time. The neuron that wins the competition is called a "winner-take-all" neuron.

Accordingly, the individual neurons of the network learn to specialize on ensembles of similar patterns and in so doing become 'feature detectors' for different classes of input patterns.

The fact that competitive networks recode sets of correlated inputs to one of a few output neurons essentially removes the redundancy in representation which is an essential part of processing in biological sensory systems.

Architecture and Implementation

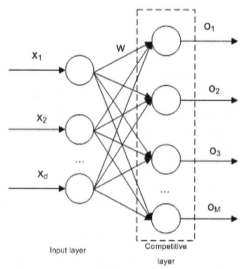

Competitive neural network architecture

Competitive Learning is usually implemented with Neural Networks that contain a hidden layer which is commonly known as "competitive layer". Every competitive neuron is described by a vector of weights $\mathbf{w}_i = \left(w_{i1}, .., w_{id} \right)^T, i = 1, .., M$ and calculates the similarity measure between the input data $\mathbf{x}^n = \left(x_{n1}, .., x_{nd} \right)^T \in \mathbb{R}^d$ and the weight vector $\mathbf{w}_i$.

For every input vector, the competitive neurons "compete" with each other to see which one of them is the most similar to that particular input vector. The winner neuron m sets its output $o_m = 1$ and all the other competitive neurons set their output $o_i = 0, i = 1, .., M, i \neq m$.

Usually, in order to measure similarity the inverse of the Euclidean distance is used: $\left\| x - w_i \right\|$ between the input vector $\mathbf{x}^n$ and the weight vector $\mathbf{w}_i$.

Example Algorithm

Here is a simple competitive learning algorithm to find three clusters within some input data.

1. (Set-up.) Let a set of sensors all feed into three different nodes, so that every node is connected to every sensor. Let the weights that each node gives to its sensors be set randomly between 0.0 and 1.0. Let the output of each node be the sum of all its sensors, each sensor's signal strength being multiplied by its weight.

2. When the net is shown an input, the node with the highest output is deemed the winner. The input is classified as being within the cluster corresponding to that node.

3. The winner updates each of its weights, moving weight from the connections that gave it weaker signals to the connections that gave it stronger signals.

Thus, as more data are received, each node converges on the centre of the cluster that it has come to represent and activates more strongly for inputs in this cluster and more weakly for inputs in other clusters.

Cluster Analysis

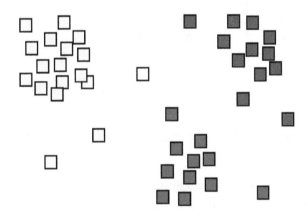

The result of a cluster analysis shown as the coloring of the squares into three clusters.

Cluster analysis or clustering is the task of grouping a set of objects in such a way that objects in the same group (called a cluster) are more similar (in some sense or another) to each other than to those in other groups (clusters). It is a main task of exploratory data mining, and a common technique for statistical data analysis, used in many fields, including machine learning, pattern recognition, image analysis, information retrieval, bioinformatics, data compression, and computer graphics.

Cluster analysis itself is not one specific algorithm, but the general task to be solved. It can be achieved by various algorithms that differ significantly in their notion of what constitutes a cluster and how to efficiently find them. Popular notions of clusters include groups with small distances among the cluster members, dense areas of the data space, intervals or particular statistical distributions. Clustering can therefore be formulated as a multi-objective optimization problem. The appropriate clustering algorithm and parameter settings (including values such as the distance function to use, a density threshold or the number of expected clusters) depend on the individual data set and intended use of the results. Cluster analysis as such is not an automatic task, but an iterative process of knowledge discovery or interactive multi-objective optimization that involves trial and failure. It is often necessary to modify data preprocessing and model parameters until the result achieves the desired properties.

Besides the term *clustering*, there are a number of terms with similar meanings, including *automatic classification, numerical taxonomy, botryology* (from Greek "grape") and *typological analysis*. The subtle differences are often in the usage of the results: while in data mining, the resulting groups are the matter of interest, in automatic classification the resulting discriminative power is of interest.

Cluster analysis was originated in anthropology by Driver and Kroeber in 1932 and introduced to psychology by Zubin in 1938 and Robert Tryon in 1939 and famously used by Cattell beginning in 1943 for trait theory classification in personality psychology.

Definition

The notion of a "cluster" cannot be precisely defined, which is one of the reasons why there are so many clustering algorithms. There is a common denominator: a group of data objects. However, different researchers employ different cluster models, and for each of these cluster models again different algorithms can be given. The notion of a cluster, as found by different algorithms, varies significantly in its properties. Understanding these "cluster models" is key to understanding the differences between the various algorithms. Typical cluster models include:

- *Connectivity models*: for example, hierarchical clustering builds models based on distance connectivity.

- *Centroid models*: for example, the k-means algorithm represents each cluster by a single mean vector.

- *Distribution models*: clusters are modeled using statistical distributions, such as multivariate normal distributions used by the expectation-maximization algorithm.

- *Density models*: for example, DBSCAN and OPTICS defines clusters as connected dense regions in the data space.

- *Subspace models*: in biclustering (also known as co-clustering or two-mode-clustering), clusters are modeled with both cluster members and relevant attributes.

- *Group models*: some algorithms do not provide a refined model for their results and just provide the grouping information.

- *Graph-based models*: a clique, that is, a subset of nodes in a graph such that every two nodes in the subset are connected by an edge can be considered as a prototypical form of cluster. Relaxations of the complete connectivity requirement (a fraction of the edges can be missing) are known as quasi-cliques, as in the HCS clustering algorithm.

A "clustering" is essentially a set of such clusters, usually containing all objects in the data set. Additionally, it may specify the relationship of the clusters to each other, for example, a hierarchy of clusters embedded in each other. Clusterings can be roughly distinguished as:

- *Hard clustering*: each object belongs to a cluster or not

- *Soft clustering* (also: *fuzzy clustering*): each object belongs to each cluster to a certain degree (for example, a likelihood of belonging to the cluster)

There are also finer distinctions possible, for example:

- *Strict partitioning clustering*: each object belongs to exactly one cluster

- *Strict partitioning clustering with outliers*: objects can also belong to no cluster, and are considered outliers

- *Overlapping clustering* (also: *alternative clustering, multi-view clustering*): objects may belong to more than one cluster; usually involving hard clusters

- *Hierarchical clustering*: objects that belong to a child cluster also belong to the parent cluster

- *Subspace clustering*: while an overlapping clustering, within a uniquely defined subspace, clusters are not expected to overlap

Algorithms

Clustering algorithms can be categorized based on their cluster model. The following overview will only list the most prominent examples of clustering algorithms, as there are possibly over 100 published clustering algorithms. Not all provide models for their clusters and can thus not easily be categorized.

There is no objectively "correct" clustering algorithm, but as it was noted, "clustering is in the eye of the beholder." The most appropriate clustering algorithm for a particular problem often needs to be chosen experimentally, unless there is a mathematical reason to prefer one cluster model over another. It should be noted that an algorithm that is designed for one kind of model will generally fail on a data set that contains a radically different kind of model. For example, k-means cannot find non-convex clusters.

Connectivity-based Clustering (Hierarchical Clustering)

Connectivity based clustering, also known as *hierarchical clustering*, is based on the core idea of objects being more related to nearby objects than to objects farther away. These algorithms connect "objects" to form "clusters" based on their distance. A cluster can be described largely by the maximum distance needed to connect parts of the cluster. At different distances, different clusters will form, which can be represented using a dendrogram, which explains where the common name "hierarchical clustering" comes from: these algorithms do not provide a single partitioning of the data set, but instead provide an extensive hierarchy of clusters that merge with each other at certain distances. In a dendrogram, the y-axis marks the distance at which the clusters merge, while the objects are placed along the x-axis such that the clusters don't mix.

Connectivity based clustering is a whole family of methods that differ by the way distances are computed. Apart from the usual choice of distance functions, the user also needs to decide on the linkage criterion (since a cluster consists of multiple objects, there are multiple candidates to compute the distance to) to use. Popular choices are known as single-linkage clustering (the minimum of object distances), complete linkage clustering (the maximum of object distances) or UPGMA ("Unweighted Pair Group Method with Arithmetic Mean", also known as average linkage clustering). Furthermore, hierarchical clustering can be agglomerative (starting with single elements and aggregating them into clusters) or divisive (starting with the complete data set and dividing it into partitions).

These methods will not produce a unique partitioning of the data set, but a hierarchy from which the user still needs to choose appropriate clusters. They are not very robust towards outliers, which will either show up as additional clusters or even cause other clusters to merge (known as "chaining phenomenon", in particular with single-linkage clustering). In the general case, the complexity is $\mathcal{O}(n^3)$ for agglomerative clustering and $\mathcal{O}(2^{n-1})$ for divisive clustering, which makes them too slow for large data sets. For some special cases, optimal efficient methods (of complexity $\mathcal{O}(n^2)$) are known: SLINK for single-linkage and CLINK for complete-linkage clustering. In the data mining community these methods are recognized as a theoretical foundation of cluster analysis, but often considered obsolete. They did however provide inspiration for many later methods such as density based clustering.

- Linkage clustering examples

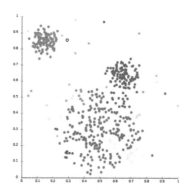

Single-linkage on Gaussian data. At 35 clusters, the biggest cluster starts fragmenting into smaller parts, while before it was still connected to the second largest due to the single-link effect.

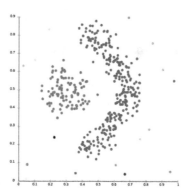

Single-linkage on density-based clusters. 20 clusters extracted, most of which contain single elements, since linkage clustering does not have a notion of "noise".

Centroid-based Clustering

In centroid-based clustering, clusters are represented by a central vector, which may not necessarily be a member of the data set. When the number of clusters is fixed to k, k-means clustering gives a formal definition as an optimization problem: find the k cluster centers and assign the objects to the nearest cluster center, such that the squared distances from the cluster are minimized.

The optimization problem itself is known to be NP-hard, and thus the common approach is to search only for approximate solutions. A particularly well known approximative method is Lloyd's algorithm, often actually referred to as "*k-means algorithm*". It does however only find a local optimum, and is commonly run multiple times with different random initializations. Variations of k-means often include such optimizations as choosing the best of multiple runs, but also restricting the centroids to members of the data set (k-medoids), choosing medians (k-medians clustering), choosing the initial centers less randomly (k-means++) or allowing a fuzzy cluster assignment (fuzzy c-means).

Most k-means-type algorithms require the number of clusters - k - to be specified in advance, which is considered to be one of the biggest drawbacks of these algorithms. Furthermore, the algorithms prefer clusters of approximately similar size, as they will always assign an object to the nearest centroid. This often leads to incorrectly cut borders in between of clusters (which is not surprising, as the algorithm optimized cluster centers, not cluster borders).

K-means has a number of interesting theoretical properties. First, it partitions the data space into a structure known as a Voronoi diagram. Second, it is conceptually close to nearest neighbor classification, and as such is popular in machine learning. Third, it can be seen as a variation of model based clustering, and Lloyd's algorithm as a variation of the Expectation-maximization algorithm for this model discussed below.

- k-means clustering examples

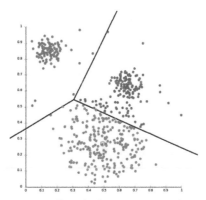

K-means separates data into Voronoi-cells, which assumes equal-sized clusters (not adequate here).

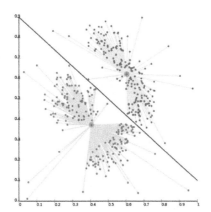

K-means cannot represent density-based clusters.

Distribution-based Clustering

The clustering model most closely related to statistics is based on distribution models. Clusters can then easily be defined as objects belonging most likely to the same distribution. A convenient property of this approach is that this closely resembles the way artificial data sets are generated: by sampling random objects from a distribution.

While the theoretical foundation of these methods is excellent, they suffer from one key problem known as overfitting, unless constraints are put on the model complexity. A more complex model will usually be able to explain the data better, which makes choosing the appropriate model complexity inherently difficult.

One prominent method is known as Gaussian mixture models (using the expectation-maximization algorithm). Here, the data set is usually modelled with a fixed (to avoid overfitting) number of Gaussian distributions that are initialized randomly and whose parameters are iteratively optimized to better fit the data set. This will converge to a local optimum, so multiple runs may produce different results. In order to obtain a hard clustering, objects are often then assigned to the Gaussian distribution they most likely belong to; for soft clusterings, this is not necessary.

Distribution-based clustering produces complex models for clusters that can capture correlation and dependence between attributes. However, these algorithms put an extra burden on the user: for many real data sets, there may be no concisely defined mathematical model (e.g. assuming Gaussian distributions is a rather strong assumption on the data).

- Expectation-maximization (EM) clustering examples

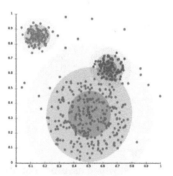

On Gaussian-distributed data, EM works well, since it uses Gaussians for modelling clusters

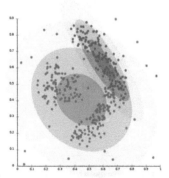

Density-based clusters cannot be modeled using Gaussian distributions

Density-based Clustering

In density-based clustering, clusters are defined as areas of higher density than the remainder of the data set. Objects in these sparse areas - that are required to separate clusters - are usually considered to be noise and border points.

The most popular density based clustering method is DBSCAN. In contrast to many newer methods, it features a well-defined cluster model called "density-reachability". Similar to linkage based clustering, it is based on connecting points within certain distance thresholds. However, it only connects points that satisfy a density criterion, in the original variant defined as a minimum number of other objects within this radius. A cluster consists of all density-connected objects (which can form a cluster of an arbitrary shape, in contrast to many other methods) plus all objects that are within these objects' range. Another interesting property of DBSCAN is that its complexity is fairly low - it requires a linear number of range queries on the database - and that it will discover essentially the same results (it is deterministic for core and noise points, but not for border points) in each run, therefore there is no need to run it multiple times. OPTICS is a generalization of DBSCAN that removes the need to choose an appropriate value for the range parameter ε, and produces a hierarchical result related to that of linkage clustering. DeLi-Clu, Density-Link-Clustering combines ideas from single-linkage clustering and OPTICS, eliminating the ε parameter entirely and offering performance improvements over OPTICS by using an R-tree index.

The key drawback of DBSCAN and OPTICS is that they expect some kind of density drop to detect cluster borders. On data sets with, for example, overlapping Gaussian distributions - a common use case in artificial data - the cluster borders produced by these algorithms will often look arbitrary, because the cluster density decreases continuously. On a data set consisting of mixtures of Gaussians, these algorithms are nearly always outperformed by methods such as EM clustering that are able to precisely model this kind of data.

Mean-shift is a clustering approach where each object is moved to the densest area in its vicinity, based on kernel density estimation. Eventually, objects converge to local maxima of density. Similar to k-means clustering, these "density attractors" can serve as representatives for the data set, but mean-shift can detect arbitrary-shaped clusters similar to DBSCAN. Due to the expensive iterative procedure and density estimation, mean-shift is usually slower than DBSCAN or k-Means. Besides that, the applicability of the mean-shift algorithm to multidimensional data is hindered by the unsmooth behaviour of the kernel density estimate, which results in over-fragmentation of cluster tails.

- Density-based clustering examples

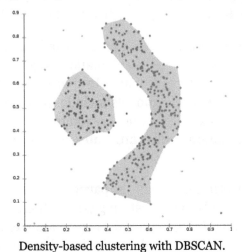

Density-based clustering with DBSCAN.

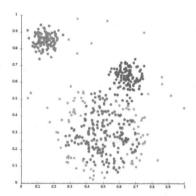

DBSCAN assumes clusters of similar density, and may have problems separating nearby clusters

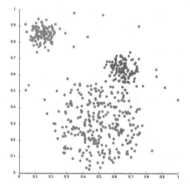

OPTICS is a DBSCAN variant that handles different densities much better

Recent Developments

In recent years considerable effort has been put into improving the performance of existing algorithms. Among them are *CLARANS* (Ng and Han, 1994), and *BIRCH* (Zhang et al., 1996). With the recent need to process larger and larger data sets (also known as big data), the willingness to trade semantic meaning of the generated clusters for performance has been increasing. This led to the development of pre-clustering methods such as canopy clustering, which can process huge data sets efficiently, but the resulting "clusters" are merely a rough pre-partitioning of the data set to then analyze the partitions with existing slower methods such as k-means clustering. Various other approaches to clustering have been tried such as seed based clustering.

For high-dimensional data, many of the existing methods fail due to the curse of dimensionality, which renders particular distance functions problematic in high-dimensional spaces. This led to new clustering algorithms for high-dimensional data that focus on subspace clustering (where only some attributes are used, and cluster models include the relevant attributes for the cluster) and correlation clustering that also looks for arbitrary rotated ("correlated") subspace clusters that can be modeled by giving a correlation of their attributes. Examples for such clustering algorithms are CLIQUE and SUBCLU.

Ideas from density-based clustering methods (in particular the DBSCAN/OPTICS family of algorithms) have been adopted to subspace clustering (HiSC, hierarchical subspace clustering and DiSH) and correlation clustering (HiCO, hierarchical correlation clustering, 4C using "correlation connectivity" and ERiC exploring hierarchical density-based correlation clusters).

Several different clustering systems based on mutual information have been proposed. One is Marina Meilă's *variation of information* metric; another provides hierarchical clustering. Using genetic algorithms, a wide range of different fit-functions can be optimized, including mutual information. Also message passing algorithms, a recent development in computer science and statistical physics, has led to the creation of new types of clustering algorithms.

Evaluation and Assessment

Evaluation (or "validation") of clustering results is as difficult as the clustering itself. Popular approaches involve *"internal"* evaluation, where the clustering is summarized to a single quality score, *"external"* evaluation, where the clustering is compared to an existing "ground truth" classification, *"manual"* evaluation by a human expert, and *"indirect"* evaluation by evaluating the utility of the clustering in its intended application.

Internal evaluation measures suffer from the problem that they represent functions that themselves can be seen as a clustering objective. For example, one could cluster the data set by the optimum Silhouette coefficient; except that there is no known efficient algorithm for this. By using such an internal measure for evaluation, we rather compare the similarity of the optimization problems, and not necessarily how useful the clustering is.

External evaluation has similar problems: if we have such "ground truth" labels, then we would not need to cluster; and in practical applications we usually do not have such labels. On the other hand, the labels only reflect one possible partitioning of the data set, which does not imply that there does not exist a different, and maybe even better, clustering.

Neither of these approaches can therefore ultimately judge the actual quality of a clustering, but this needs human evaluation, which is highly subjective. Nevertheless, such statistics can be quite informative in identifying bad clusterings, but one should not dismiss subjective human evaluation.

Internal Evaluation

When a clustering result is evaluated based on the data that was clustered itself, this is called internal evaluation. These methods usually assign the best score to the algorithm that produces clusters with high similarity within a cluster and low similarity between clusters. One drawback of using internal criteria in cluster evaluation is that high scores on an internal measure do not necessarily result in effective information retrieval applications. Additionally, this evaluation is biased towards algorithms that use the same cluster model. For example, k-means clustering naturally optimizes object distances, and a distance-based internal criterion will likely overrate the resulting clustering.

Therefore, the internal evaluation measures are best suited to get some insight into situations where one algorithm performs better than another, but this shall not imply that one algorithm produces more valid results than another. Validity as measured by such an index depends on the claim that this kind of structure exists in the data set. An algorithm designed for some kind of models has no chance if the data set contains a radically different set of models, or if the evaluation measures a radically different criterion. For example, k-means clustering can only find convex clusters, and many evaluation indexes assume convex clusters. On a data set with non-convex clusters neither the use of k-means, nor of an evaluation criterion that assumes convexity, is sound.

The following methods can be used to assess the quality of clustering algorithms based on internal criterion:

- Davies–Bouldin index

  The Davies–Bouldin index can be calculated by the following formula:

  $$DB = \frac{1}{n} \sum_{i=1}^{n} \max_{j \neq i} \left(\frac{\sigma_i + \sigma_j}{d(c_i, c_j)} \right)$$

  where n is the number of clusters, c_x is the centroid of cluster x, σ_x is the average distance of all elements in cluster x to centroid c_x, and $d(c_i, c_j)$ is the distance between centroids c_i and c_j. Since algorithms that produce clusters with low intra-cluster distances (high intra-cluster similarity) and high inter-cluster distances (low inter-cluster similarity) will have a low Davies–Bouldin index, the clustering algorithm that produces a collection of clusters with the smallest Davies–Bouldin index is considered the best algorithm based on this criterion.

- Dunn index

  The Dunn index aims to identify dense and well-separated clusters. It is defined as the ratio between the minimal inter-cluster distance to maximal intra-cluster distance. For each cluster partition, the Dunn index can be calculated by the following formula:

  $$D = \frac{\min_{1 \leq i < j \leq n} d(i, j)}{\max_{1 \leq k \leq n} d'(k)},$$

  where $d(i,j)$ represents the distance between clusters i and j, and $d'(k)$ measures the intra-cluster distance of cluster k. The inter-cluster distance $d(i,j)$ between two clusters may be any number of distance measures, such as the distance between the centroids of the clusters. Similarly, the intra-cluster distance $d'(k)$ may be measured in a variety ways, such as the maximal distance between any pair of elements in cluster k. Since internal criterion seek clusters with high intra-cluster similarity and low inter-cluster similarity, algorithms that produce clusters with high Dunn index are more desirable.

- Silhouette coefficient

  The silhouette coefficient contrasts the average distance to elements in the same cluster with the average distance to elements in other clusters. Objects with a high silhouette value are considered well clustered, objects with a low value may be outliers. This index works well with k-means clustering, and is also used to determine the optimal number of clusters.

External Evaluation

In external evaluation, clustering results are evaluated based on data that was not used for clustering, such as known class labels and external benchmarks. Such benchmarks consist of a set of pre-classified items, and these sets are often created by (expert) humans. Thus, the benchmark sets can be thought of as a gold standard for evaluation. These types of evaluation methods measure how close the clustering is to the predetermined benchmark classes. However, it has recently been discussed whether this is adequate for real data, or only on synthetic data sets with a factual ground truth, since classes can contain internal structure, the attributes present may not allow separation of clusters or the classes may contain anomalies. Additionally, from a knowledge discovery point of view, the reproduction of known knowledge may not necessarily be the intended result. In the special scenario of constrained clustering, where meta information (such as class labels) is used already in the clustering process, the hold-out of information for evaluation purposes is non-trivial.

A number of measures are adapted from variants used to evaluate classification tasks. In place of counting the number of times a class was correctly assigned to a single data point (known as true positives), such *pair counting* metrics assess whether each pair of data points that is truly in the same cluster is predicted to be in the same cluster.

Some of the measures of quality of a cluster algorithm using external criterion include:

- Purity: Purity is a measure of the extent to which clusters contain a single class. Its calculation can be thought of as follows: For each cluster, count the number of data points from the most common class in said cluster. Now take the sum over all clusters and divide by the total number of data points. Formally, given some set of clusters M and some set of classes D, both partitioning N data points, purity can be defined as:

$$\frac{1}{N} \sum_{m \in M} \max_{d \in D} |m \cap d|$$

  Note that this measure doesn't penalise having many clusters. So for example, a purity score of 1 is possible by putting each data point in its own cluster.

- Rand measure (William M. Rand 1971)

  The Rand index computes how similar the clusters (returned by the clustering algorithm) are to the benchmark classifications. One can also view the Rand index as a measure of the percentage of correct decisions made by the algorithm. It can be computed using the following formula:

$$RI = \frac{TP + TN}{TP + FP + FN + TN}$$

  where TP is the number of true positives, TN is the number of true negatives, FP is the number of false positives, and FN is the number of false negatives. One issue with the Rand index is that false positives and false negatives are equally weighted. This may be an undesirable characteristic for some clustering applications. The F-measure addresses this concern, as does the chance-corrected adjusted Rand index.

- F-measure

  The F-measure can be used to balance the contribution of false negatives by weighting recall through a parameter $\beta \geq 0$. Let precision and recall (both external evaluation measures in themselves) be defined as follows:

$$P = \frac{TP}{TP + FP}$$

$$R = \frac{TP}{TP + FN}$$

where P is the precision rate and R is the recall rate. We can calculate the F-measure by using the following formula:

$$F_\beta = \frac{(\beta^2 + 1) \cdot P \cdot R}{\beta^2 \cdot P + R}$$

Notice that when $\beta = 0$, $F_0 = P$. In other words, recall has no impact on the F-measure when $\beta = 0$, and increasing β allocates an increasing amount of weight to recall in the final F-measure.

- Jaccard index

The Jaccard index is used to quantify the similarity between two datasets. The Jaccard index takes on a value between 0 and 1. An index of 1 means that the two dataset are identical, and an index of 0 indicates that the datasets have no common elements. The Jaccard index is defined by the following formula:

$$J(A,B) = \frac{|A \cap B|}{|A \cup B|} = \frac{TP}{TP + FP + FN}$$

This is simply the number of unique elements common to both sets divided by the total number of unique elements in both sets.

- Fowlkes–Mallows index (E. B. Fowlkes & C. L. Mallows 1983)

The Fowlkes-Mallows index computes the similarity between the clusters returned by the clustering algorithm and the benchmark classifications. The higher the value of the Fowlkes-Mallows index the more similar the clusters and the benchmark classifications are. It can be computed using the following formula:

$$FM = \sqrt{\frac{TP}{TP + FP} \cdot \frac{TP}{TP + FN}}$$

where TP is the number of true positives, FP is the number of false positives, and FN is the number of false negatives. The FM index is the geometric mean of the precision and recall P and R, while the F-measure is their harmonic mean. Moreover, precision and recall are also known as Wallace's indices B^I and B^{II}.

- The mutual information is an information theoretic measure of how much information is shared between a clustering and a ground-truth classification that can detect a non-linear similarity between two clusterings. Adjusted mutual information is the corrected-for-chance variant of this that has a reduced bias for varying cluster numbers.

- Confusion matrix

  A confusion matrix can be used to quickly visualize the results of a classification (or clustering) algorithm. It shows how different a cluster is from the gold standard cluster.

Cluster Tendency

To measure cluster tendency is to measure to what degree clusters exist in the data to be clustered, and may be performed as an initial test, before attempting clustering. One way to do this is to compare the data against random data. On average, random data should not have clusters.

- Hopkins statistic

  There are multiple formulations of the Hopkins Statistic. A typical one is as follows. Let X be the set of n data points in d dimensional space. Consider a random sample (without replacement) of $m \ll n$ data points with members x_i. Also generate a set Y of m uniformly randomly distributed data points. Now define two distance measures, u_i to be the distance of $y_i \in Y$ from its nearest neighbor in X and w_i to be the distance of $x_i \in X$ from its nearest neighbor in X. We then define the Hopkins statistic as:

  $$H = \frac{\sum_{i=1}^{m} u_i^d}{\sum_{i=1}^{m} u_i^d + \sum_{i=1}^{m} w_i^d},$$

  With this definition, uniform random data should tend to have values near to 0.5, and clustered data should tend to have values nearer to 1.

  However, data containing just a single Gaussian will also score close to 1, as this statistic measures deviation from a *uniform* distribution, not multimodality, making this statistic largely useless in application (as real data never is remotely uniform).

Applications

Biology, computational biology and bioinformatics

  Plant and animal ecology

  cluster analysis is used to describe and to make spatial and temporal comparisons of communities (assemblages) of organisms in heterogeneous environments; it is also used in plant systematics to generate artificial phylogenies or clusters of organisms (individuals) at the species, genus or higher level that share a number of attributes

  Transcriptomics

  clustering is used to build groups of genes with related expression patterns (also known as coexpressed genes) as in HCS clustering algorithm. Often such groups contain functionally related proteins, such as enzymes for a specific pathway, or genes that are co-regulated. High throughput experiments using expressed sequence tags (ESTs) or DNA microarrays can be a powerful tool for genome annotation, a general aspect of genomics.

Sequence analysis

clustering is used to group homologous sequences into gene families. This is a very important concept in bioinformatics, and evolutionary biology in general.

High-throughput genotyping platforms

clustering algorithms are used to automatically assign genotypes.

Human genetic clustering

The similarity of genetic data is used in clustering to infer population structures.

Medicine

Medical imaging

On PET scans, cluster analysis can be used to differentiate between different types of tissue in a three-dimensional image for many different purposes.

Analysis of antimicrobial activity

Cluster analysis can be used to analyse patterns of antibiotic resistance, to classify antimicrobial compounds according to their mechanism of action, to classify antibiotics according to their antibacterial activity.

IMRT segmentation

Clustering can be used to divide a fluence map into distinct regions for conversion into deliverable fields in MLC-based Radiation Therapy.

Business and marketing

Market research

Cluster analysis is widely used in market research when working with multivariate data from surveys and test panels. Market researchers use cluster analysis to partition the general population of consumers into market segments and to better understand the relationships between different groups of consumers/potential customers, and for use in market segmentation, Product positioning, New product development and Selecting test markets.

Grouping of shopping items

Clustering can be used to group all the shopping items available on the web into a set of unique products. For example, all the items on eBay can be grouped into unique products. (eBay doesn't have the concept of a SKU)

World wide web

Social network analysis

In the study of social networks, clustering may be used to recognize communities within large groups of people.

Search result grouping

In the process of intelligent grouping of the files and websites, clustering may be used to create a more relevant set of search results compared to normal search engines like Google. There are currently a number of web based clustering tools such as Clusty.

Slippy map optimization

Flickr's map of photos and other map sites use clustering to reduce the number of markers on a map. This makes it both faster and reduces the amount of visual clutter.

Computer science

Software evolution

Clustering is useful in software evolution as it helps to reduce legacy properties in code by reforming functionality that has become dispersed. It is a form of restructuring and hence is a way of direct preventative maintenance.

Image segmentation

Clustering can be used to divide a digital image into distinct regions for border detection or object recognition.

Evolutionary algorithms

Clustering may be used to identify different niches within the population of an evolutionary algorithm so that reproductive opportunity can be distributed more evenly amongst the evolving species or subspecies.

Recommender systems

Recommender systems are designed to recommend new items based on a user's tastes. They sometimes use clustering algorithms to predict a user's preferences based on the preferences of other users in the user's cluster.

Markov chain Monte Carlo methods

Clustering is often utilized to locate and characterize extrema in the target distribution.

Anomaly detection

Anomalies/outliers are typically - be it explicitly or implicitly - defined with respect to clustering structure in data.

Social science

Crime analysis

Cluster analysis can be used to identify areas where there are greater incidences of particular types of crime. By identifying these distinct areas or "hot spots" where a similar crime has happened over a period of time, it is possible to manage law enforcement resources more effectively.

Educational data mining

Cluster analysis is for example used to identify groups of schools or students with similar properties.

Typologies

From poll data, projects such as those undertaken by the Pew Research Center use cluster analysis to discern typologies of opinions, habits, and demographics that may be useful in politics and marketing.

Others

Field robotics

Clustering algorithms are used for robotic situational awareness to track objects and detect outliers in sensor data.

Mathematical chemistry

To find structural similarity, etc., for example, 3000 chemical compounds were clustered in the space of 90 topological indices.

Climatology

To find weather regimes or preferred sea level pressure atmospheric patterns.

Petroleum geology

Cluster analysis is used to reconstruct missing bottom hole core data or missing log curves in order to evaluate reservoir properties.

Physical geography

The clustering of chemical properties in different sample locations.

Generative Adversarial Networks

Generative adversarial networks (GANs) are a type of artificial intelligence algorithms used in unsupervised machine learning, implemented by a system of two neural networks competing against each other in a zero-sum game framework. They were first introduced by Ian Goodfellow *et al.* in 2014.

This technique can generate photographs that look authentic to human observers.

Method

One network is generative and one is discriminative. Typically, the generative network is taught to map from a latent space to a particular data distribution of interest, and the discriminative network is simultaneously taught to discriminate between instances from the true data distribution and synthesized instances produced by the generator. The generative network's training objective is to increase the error rate of the discriminative network (i.e., "fool" the discriminator network by producing novel synthesized instances that appear to have come from the true data distribution). These models are used for computer vision tasks.

In practice, a particular dataset serves as the training data for the discriminator. Training the discriminator involves presenting the discriminator with samples from the dataset and samples synthesized by the generator, and backpropagating from a binary classification loss. In order to produce a sample, typically the generator is seeded with a randomized input that is sampled from a predefined latent space (e.g., a multivariate normal distribution). Training the generator involves back-propagating the negation of the binary classification loss of the discriminator. The generator adjusts its parameters so that the training data and generated data cannot be distinguished by the discriminator model. The goal is to find a setting of parameters that makes generated data look like the training data to the discriminator network. In practice, the generator is typically a deconvolutional neural network, and the discriminator is a convolutional neural network.

The idea to infer models in a competitive setting (model versus discriminator) was first proposed by Li, Gauci and Gross in 2013. Their method is used for behavioral inference. It is termed Turing Learning, as the setting is akin to that of a Turing test.

Application

GANs can be used to produce samples of photorealistic images for the purposes of visualizing new interior/industrial design, shoes, bags and clothing items or items for computer games' scenes. These networks were reported to be used by Facebook. Recently, GANs have been able to model rudimentary patterns of motion in video. They have also been used to reconstruct 3D models of objects from images and to improve astronomical images.

Autoencoder

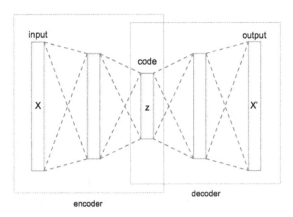

Schematic structure of an autoencoder with 3 fully-connected hidden layers.

An autoencoder, autoassociator or Diabolo network is an artificial neural network used for unsupervised learning of efficient codings. The aim of an autoencoder is to learn a representation (encoding) for a set of data, typically for the purpose of dimensionality reduction. Recently, the autoencoder concept has become more widely used for learning generative models of data.

Structure

Architecturally, the simplest form of an autoencoder is a feedforward, non-recurrent neural network very similar to the multilayer perceptron (MLP) – having an input layer, an output layer and one or more hidden layers connecting them –, but with the output layer having the same number of nodes as the input layer, and with the purpose of *reconstructing* its own inputs (instead of predicting the target value Y given inputs X). Therefore, autoencoders are unsupervised learning models.

An autoencoder always consists of two parts, the encoder and the decoder, which can be defined as transitions ϕ and ψ, such that:

$$\phi : \mathcal{X} \to \mathcal{F}$$

$$\psi : \mathcal{F} \to \mathcal{X}$$

$$\phi, \psi = \arg\min_{\phi, \psi} \| X - (\psi \circ \phi) X \|^2$$

In the simplest case, where there is one hidden layer, the encoder stage of an autoencoder takes the input $\mathbf{x} \in \mathbb{R}^d = \mathcal{X}$ and maps it to $\mathbf{z} \in \mathbb{R}^p = \mathcal{F}$:

$$\mathbf{z} = \sigma(\mathbf{Wx} + \mathbf{b})$$

This image $\mathbf{z}$ is usually referred to as *code*, *latent variables*, or *latent representation*. Here, σ is an element-wise activation function such as a sigmoid function or a rectified linear unit. $\mathbf{W}$ is a weight matrix and $\mathbf{b}$ is a bias vector. After that, the decoder stage of the autoencoder maps $\mathbf{z}$ to the *reconstruction* $\mathbf{x}'$ of the same shape as $\mathbf{x}$:

$$\mathbf{x}' = \sigma'(\mathbf{W}'\mathbf{z} + \mathbf{b}')$$

where $\sigma', W',$ and b' for the decoder may differ in general from the corresponding $\sigma', W',$ and b' for the encoder, depending on the design of the autoencoder.

Autoencoders are also trained to minimise reconstruction errors (such as squared errors):

$$\mathcal{L}(\mathbf{x}, \mathbf{x}') = \| \mathbf{x} - \mathbf{x}' \|^2 = \| \mathbf{x} - \sigma'(\mathbf{W}'(\sigma(\mathbf{Wx} + \mathbf{b})) + \mathbf{b}') \|^2$$

where $\mathbf{x}$ is usually averaged over some input training set.

If the feature space $\mathcal{F}$ has lower dimensionality than the input space $\mathcal{X}$, then the feature vector $\phi(x)$ can be regarded as a compressed representation of the input x. If the hidden layers are larger than the input layer, an autoencoder can potentially learn the identity function and become useless. However, experimental results have shown that autoencoders might still learn useful features in these cases.

Variations

Various techniques exist to prevent autoencoders from learning the identity function and to improve their ability to capture important information and learn richer representations:

Denoising Autoencoder

Denoising autoencoders take a partially corrupted input whilst training to recover the original un-distorted input. This technique has been introduced with a specific approach to *good* representation. A *good representation is one that can be obtained robustly from a corrupted input and that will be useful for recovering the corresponding clean input.* This definition contains the following implicit assumptions:

- The higher level representations are relatively stable and robust to the corruption of the input;

- It is necessary to extract features that are useful for representation of the input distribution.

To train an autoencoder to denoise data, it is necessary to perform preliminary stochastic mapping $\mathbf{x} \rightarrow \tilde{\mathbf{x}}$ in order to corrupt the data and use $\tilde{\mathbf{x}}$ as input for a normal autoencoder, with the only exception being that the loss should be still computed for the initial input $\mathcal{L}(\mathbf{x}, \tilde{\mathbf{x}})$ instead of $\mathcal{L}(\tilde{\mathbf{x}}, \tilde{\mathbf{x}}')$.

Sparse Autoencoder

By imposing sparsity on the hidden units during training (whilst having a larger number of hidden units than inputs), an autoencoder can learn useful structures in the input data. This allows sparse representations of inputs. These are useful in pretraining for classification tasks.

Sparsity may be achieved by additional terms in the loss function during training (by comparing the probability distribution of the hidden unit activations with some low desired value), or by manually zeroing all but the few strongest hidden unit activations (referred to as a *k-sparse auto-encoder*).

Variational Autoencoder (VAE)

Variational autoencoder models inherit autoencoder architecture, but make strong assumptions concerning the distribution of latent variables. They use variational approach for latent representation learning, which results in an additional loss component and specific training algorithm called *Stochastic Gradient Variational Bayes (SGVB)*. It assumes that the data is generated by a directed graphical model $p(\mathbf{x} \mid \mathbf{z})$ and that the encoder is learning an approximation $q_\phi(\mathbf{z} \mid \mathbf{x})$ to the posterior distribution $p_\theta(\mathbf{z} \mid \mathbf{x})$ where ϕ and θ denote the parameters of the encoder (recognition model) and decoder (generative model) respectively. The objective of the variational autoencoder in this case has the following form:

$$\mathcal{L}(\phi, \theta, \mathbf{x}) = D_{KL}(q_\phi(\mathbf{z} \mid \mathbf{x}) \| p_\theta(\mathbf{z})) - \mathbb{E}_{q_\phi(\mathbf{z} \mid \mathbf{x})}\left(\log p_\theta(\mathbf{x} \mid \mathbf{z})\right)$$

Here, D_{KL} stands for the Kullback–Leibler divergence. The prior over the latent variables is usually set to be the centred isotropic multivariate Gaussian $p_\theta(\mathbf{z}) = \mathcal{N}(\mathbf{0}, \mathbf{I})$; however, alternative configurations have also been recently considered, e.g.

Contractive Autoencoder (CAE)

Contractive autoencoder adds an explicit regularizer in their objective function that forces the model to learn a function that is robust to slight variations of input values. This regularizer corresponds to the Frobenius norm of the Jacobian matrix of the encoder activations with respect to the input. The final objective function has the following form:

$$\mathcal{L}(\mathbf{x}, \mathbf{x}') + \lambda \sum_i || \nabla_x \mathbf{h}_i ||^2$$

Relationship with Truncated Singular Value Decomposition (TSVD)

If linear activations are used, or only a single sigmoid hidden layer, then the optimal solution to an autoencoder is strongly related to principal component analysis (PCA).

Training

The training algorithm for an autoencoder can be summarized as

> For each input x,
>
> Do a feed-forward pass to compute activations at all hidden layers, then at the output layer to obtain an output $\mathbf{x}'$
>
> Measure the deviation of $\mathbf{x}'$ from the input $\mathbf{x}$ (typically using squared error),
>
> Backpropagate the error through the net and perform weight updates.

An autoencoder is often trained using one of the many variants of backpropagation (such as conjugate gradient method, steepest descent, etc.). Though these are often reasonably effective, there are fundamental problems with the use of backpropagation to train networks with many hidden layers. Once errors are backpropagated to the first few layers, they become minuscule and insignificant. This means that the network will almost always learn to reconstruct the average of all the training data. Though more advanced backpropagation methods (such as the conjugate gradient method) can solve this problem to a certain extent, they still result in a very slow learning process and poor solutions. This problem can be remedied by using initial weights that approximate the final solution. The process of finding these initial weights is often referred to as *pretraining*.

Geoffrey Hinton developed a pretraining technique for training many-layered "deep" autoencoders. This method involves treating each neighbouring set of two layers as a restricted Boltzmann machine so that the pretraining approximates a good solution, then using a backpropagation technique to fine-tune the results. This model takes the name of deep belief network.

References

- M. Belkin; P. Niyogi (2004). "Semi-supervised Learning on Riemannian Manifolds". Machine Learning. 56 (Special Issue on Clustering): 209–239. doi:10.1023/b:mach.0000033120.25363.1

- Salatas, John (24 August 2011). "Implementation of Competitive Learning Networks for WEKA". ICT Research Blog. Retrieved 28 January 2012

- Mehryar Mohri, Afshin Rostamizadeh, Ameet Talwalkar (2012) Foundations of Machine Learning, The MIT Press ISBN 9780262018258

- C.E. Brodely and M.A. Friedl (1999). Identifying and Eliminating Mislabeled Training Instances, Journal of Artificial Intelligence Research 11, 131-167

- Younger B. A.; Fearing D. D. (1999). "Parsing Items into Separate Categories: Developmental Change in Infant Categorization". Child Development. 70: 291–303. doi:10.1111/1467-8624.00022

- Chapelle, Olivier; Schölkopf, Bernhard; Zien, Alexander (2006). Semi-supervised learning. Cambridge, Mass.: MIT Press. ISBN 978-0-262-03358-9

- Basak, S.C.; Magnuson, V.R.; Niemi, C.J.; Regal, R.R. (1988). "Determining Structural Similarity of Chemicals Using Graph Theoretic Indices". Discr. Appl. Math. 19: 17–44. doi:10.1016/0166-218x(88)90004-2

- Fuhr, Norbert (1992), "Probabilistic Models in Information Retrieval", Computer Journal, 35 (3): 243–255, doi:10.1093/comjnl/35.3.243

- Shakhnarovich, Greg (2011). "Notes on derivation of bias-variance decomposition in linear regression" (PDF). Archived from the original (PDF) on 21 August 2014. Retrieved 20 August 2014

- Zhu, Xiaojin; Goldberg, Andrew B. (2009). Introduction to semi-supervised learning. Morgan & Claypool. ISBN 9781598295481

- Cattell, R. B. (1943). "The description of personality: Basic traits resolved into clusters". Journal of Abnormal and Social Psychology. 38 (4): 476–506. doi:10.1037/h0054116

- Xu, F. & Tenenbaum, J. B. (2007). "Sensitivity to sampling in Bayesian word learning. Developmental Science". Developmental Science. 10: 288–297. doi:10.1111/j.1467-7687.2007.00590.x

- Vijayakumar, Sethu (2007). "The Bias–Variance Tradeoff" (PDF). University Edinburgh. Retrieved 19 August 2014

- Belsley, David (1991). Conditioning diagnostics : collinearity and weak data in regression. New York: Wiley. ISBN 978-0471528890

- Sibson, R. (1973). "SLINK: an optimally efficient algorithm for the single-link cluster method" (PDF). The Computer Journal. British Computer Society. 16 (1): 30–34. doi:10.1093/comjnl/16.1.30

- Fuhr, Norbert (1989), "Optimum polynomial retrieval functions based on the probability ranking principle", ACM Transactions on Information Systems, 7 (3): 183–204, doi:10.1145/65943.65944

- Bishop, Christopher M. "Chapter 4. Linear Models for Classification". Pattern Recognition and Machine Learning. Springer Science+Business Media, LLC. p. 194. ISBN 978-0387-31073-2

- "How Bloomberg Integrated Learning-to-Rank into Apache Solr | Tech at Bloomberg". Tech at Bloomberg. 2017-01-23. Retrieved 2017-02-28

- Geman, Stuart; E. Bienenstock; R. Doursat (1992). "Neural networks and the bias/variance dilemma" (PDF). Neural Computation. 4: 1–58. doi:10.1162/neco.1992.4.1.1

- Feldman, Ronen; Sanger, James (2007-01-01). The Text Mining Handbook: Advanced Approaches in Analyzing Unstructured Data. Cambridge Univ. Press. ISBN 0521836573. OCLC 915286380

- Wallace, D. L. (1983). "Comment". Journal of the American Statistical Association. 78 (383): 569–579. doi:10.1080/01621459.1983.10478009

- Freund, Y.; Schapire, R. E. (1999). "Large margin classification using the perceptron algorithm" (PDF). Machine Learning. 37 (3): 277–296. doi:10.1023/A:1007662407062

Methods of Machine Learning

The methods of machine learning are logic learning machine, online machine learning, rule-based machine learning, multiple kernel learning and temporal difference learning. Logic learning machine is a method which has its basis on intelligible rules. This chapter discusses the methods of machine learning in a critical manner providing key analysis to the subject matter.

Logic Learning Machine

Logic Learning Machine (LLM) is a machine learning method based on the generation of intelligible rules. LLM is an efficient implementation of the Switching Neural Network (SNN) paradigm, developed by Marco Muselli, Senior Researcher at the Italian National Research Council CNR-IEIIT in Genoa. Logic Learning Machine is implemented in the Rulex suite.

LLM has been employed in different fields, including orthopaedic patient classification, DNA microarray analysis and Clinical Decision Support System.

History

The Switching Neural Network approach was developed in the 1990s to overcome the drawbacks of the most commonly used machine learning methods. In particular, black box methods, such as multilayer perceptron and support vector machine, had good accuracy but could not provide deep insight into the studied phenomenon. On the other hand, decision trees were able to describe the phenomenon but often lacked accuracy. Switching Neural Networks made use of Boolean algebra to build sets of intelligible rules able to obtain very good performance. In 2014, an efficient version of Switching Neural Network was developed and implemented in the Rulex suite with the name Logic Learning Machine. Also a LLM version devoted to regression problems was developed.

General

Like other machine learning methods, LLM uses data to build a model able to perform a good forecast about future behaviors. LLM starts from a table including a target variable (output) and some inputs and generates a set of rules that return the output value y corresponding to a given configuration of inputs. A rule is written in the form:

$$\textbf{if } premise \textbf{ then } consequence$$

where *consequence* contains the output value whereas *premise* includes one or more conditions on the inputs. According to the input type, conditions can have different forms:

- for categorical variables the input value must be in a given subset : $x_1 \in \{A, B, C, ...\}$.

- for ordered variables the condition is written as an inequality or an interval: $x_2 \leq \alpha$ or $\beta \leq x_3 \leq \gamma$

A possible rule is therefore in the form

$$\textbf{if } x_1 \in \{A, B, C, ...\} \textbf{ AND } x_2 \leq \alpha \textbf{ AND } \beta \leq x_3 \leq \gamma \textbf{ then } y = \overline{y}$$

Types

According to the output type, different versions of Logic Learning Machine have been developed:

- Logic Learning Machine for classification, when the output is a categorical variable, which can assume values in a finite set

- Logic Learning Machine for regression, when the output is an integer or real number

Online Machine Learning

In computer science, online machine learning is a method of machine learning in which data becomes available in a sequential order and is used to update our best predictor for future data at each step, as opposed to batch learning techniques which generate the best predictor by learning on the entire training data set at once. Online learning is a common technique used in areas of machine learning where it is computationally infeasible to train over the entire dataset, requiring the need of out-of-core algorithms. It is also used in situations where it is necessary for the algorithm to dynamically adapt to new patterns in the data, or when the data itself is generated as a function of time, e.g. stock price prediction. Online learning algorithms may be prone to catastrophic interference. This problem is tackled by incremental learning approaches.

Introduction

In the setting of supervised learning, a function of $f : X \rightarrow Y$ is to be learned, where X is thought of as a space of inputs and Y as a space of outputs, that predicts well on instances that are drawn from a joint probability distribution $p(x, y)$ on $X \times Y$. In reality, the learner never knows the true distribution $p(x, y)$ over instances. Instead, the learner usually has access to a training set of examples $(x_1, y_1), ..., (x_n, y_n)$. In this setting, the loss function is given as $V : Y \times Y \rightarrow \mathbb{R}$, such that $V(f(x), y)$ measures the difference between the predicted value $f(x)$ and the true value y. The ideal goal is to select a function $f \in \mathcal{H}$, where $\mathcal{H}$ is a space of functions called a hypothesis space, so that some notion of total loss is minimised. Depending on the type of model (statistical or adversarial), one can devise different notions of loss, which lead to different learning algorithms.

Statistical View of Online Learning

In statistical learning models, the training sample (x_i, y_i) are assumed to have been drawn from the true distribution $p(x, y)$ and the objective is to minimize the expected "risk"

$$I[f] = \mathbb{E}[V(f(x), y)] = \int V(f(x), y) dp(x, y).$$

A common paradigm in this situation is to estimate a function $\hat{f}$ through empirical risk minimization or regularized empirical risk minimization (usually Tikhonov regularization). The choice of loss function here gives rise to several well-known learning algorithms such as regularized least squares and support vector machines. A purely online model in this category would learn based on just the new input (x_{t+1}, y_{t+1}), the current best predictor f_t and some extra stored information (which is usually expected to have storage requirements independent of training data size). For many formulations, for example nonlinear kernel methods, true online learning is not possible, though a form of hybrid online learning with recursive algorithms can be used where f_{t+1} is permitted to depend on f_t and all previous data points $(x_1, y_1), \ldots, (x_t, y_t)$. In this case, the space requirements are no longer guaranteed to be constant since it requires storing all previous data points, but the solution may take less time to compute with the addition of a new data point, as compared to batch learning techniques.

A common strategy to overcome the above issues is to learn using mini-batches, which process a small batch of $b \geq 1$ data points at a time, this can be considered as pseudo-online learning for b much smaller than the total number of training points. Mini-batch techniques are used with repeated passing over the training data to obtain optimized out-of-core versions of machine learning algorithms, for e.g. Stochastic gradient descent. When combined with backpropagation, this is currently the de facto training method for training artificial neural networks.

Example: Linear Least Squares

The simple example of linear least squares is used to explain a variety of ideas in online learning. The ideas are general enough to be applied to other settings, for e.g. with other convex loss functions.

Batch Learning

In the setting of supervised learning with the square loss function, the intent is to minimize the empirical loss,

$$I_n[w] = \sum_{j=1}^{n} V(\langle w, x_j \rangle, y_j) = \sum_{j=1}^{n} (x_j^T w - y_j)^2 \text{ where}$$

$$x_j \in \mathbb{R}^d, w \in \mathbb{R}^d, y_j \in \mathbb{R}.$$

Let X be the $i \times d$ data matrix and Y is the $i \times 1$ matrix of target values after the arrival of the first i data points. Assuming that the covariance matrix $\Sigma_i = X^T X$ is invertible (otherwise it is preferential to proceed in a similar fashion with Tikhonov regularization), the best solution $f^*(x) = \langle w^*, x \rangle$

to the linear least squares problem is given by

$$w^* = (X^TX)^{-1}X^TY = \Sigma_i^{-1}\sum_{j=1}^{i} x_j y_j.$$

Now, calculating the covariance matrix $\Sigma_i = \sum_{j=1}^{i} x_j x_j^T$ takes time $O(id^2)$, inverting the $d \times d$ matrix takes time $O(d^3)$, while the rest of the multiplication takes time $O(d^2)$, giving a total time of $O(id^2 + d^3)$. When n total points in the dataset and having to recompute the solution after the arrival of every datapoint $i = 1,\ldots,n$, the naive approach will have a total complexity $O(n^2d^2 + nd^3)$. Note that when storing the matrix Σ_i, then updating it at each step needs only adding $x_{i+1}x_{i+1}^T$, which takes $O(d^2)$ time, reducing the total time to $O(nd^2 + nd^3) = O(nd^3)$, but with an additional storage space of $O(d^2)$ to store Σ_i.

Online Learning: Recursive Least Squares

The recursive least squares algorithm considers an online approach to the least squares problem. It can be shown that by initialising $w_0 = 0 \in \mathbb{R}^d$ and $\Gamma_0 = I \in \mathbb{R}^{d \times d}$, the solution of the linear least squares problem given can be computed by the following iteration:

$$\Gamma_i = \Gamma_{i-1} - \frac{\Gamma_{i-1}x_i x_i^T \Gamma_{i-1}}{1 + x_i^T \Gamma_{i-1} x_i}$$

$$w_i = w_{i-1} - \Gamma_i x_i (x_i^T w_{i-1} - y_i)$$

The above iteration algorithm can be proved using induction on i. The proof also shows that $\Gamma_i = \Sigma_i^{-1}$. One can look at RLS also in the context of adaptive filters.

The complexity for n steps of this algorithm is $O(nd^2)$, which is an order of magnitude faster than the corresponding batch learning complexity. The storage requirements at every step i here are to store the matrix Γ_i, which is constant at $O(d^2)$. For the case when Σ_i is not invertible, consider the regularised version of the problem loss function $\sum_{j=1}^{n}(x_j^T w - y_j)^2 + \lambda \| w \|_2^2$. Then, it's easy to show that the same algorithm works with $\Gamma_0 = (I + \lambda I)^{-1}$, and the iterations proceed to give $\Gamma_i = (\Sigma_i + \lambda I)^{-1}$.

Stochastic Gradient Descent

When this is replaced,

$$w_i = w_{i-1} - \Gamma_i x_i (x_i^T w_{i-1} - y_i)$$

by

$$w_i = w_{i-1} - \gamma_i x_i (x_i^T w_{i-1} - y_i) = w_{i-1} - \gamma_i \nabla V(\langle w_{i-1}, x_i, y_i \rangle)$$

or $\Gamma_i \in \mathbb{R}^{d \times d}$ by $\gamma_i \in \mathbb{R}$, this becomes the stochastic gradient descent algorithm. In this case, the complexity for n steps of this algorithm reduces to $O(nd)$. The storage requirements at every step i are constant at $O(d)$.

However, the stepsize γ_i needs to be chosen carefully to solve the expected risk minimization problem, as detailed above. By choosing a decaying step size $\gamma_i \approx \dfrac{1}{\sqrt{i}}$, one can prove the convergence of the average iterate $\overline{w}_n = \dfrac{1}{n}\sum_{i=1}^{n} w_i$. This setting is a special case of stochastic optimization, a well known problem in optimization.

Incremental Stochastic Gradient Descent

In practice, one can perform multiple stochastic gradient passes (also called cycles or epochs) over the data. The algorithm thus obtained is called incremental gradient method and corresponds to an iteration.

$$w_i = w_{i-1} - \gamma_i \nabla V(\langle w_{i-1}, x_{t_i} \rangle, y_{t_i})$$

The main difference with the stochastic gradient method is that here a sequence t_i is chosen to decide which training point is visited in the i-th step. Such a sequence can be stochastic or deterministic. The number of iterations is then decoupled to the number of points (each point can be considered more than once). The incremental gradient method can be shown to provide a minimizer to the empirical risk. Incremental techniques can be advantageous when considering objective functions made up of a sum of many terms e.g. an empirical error corresponding to a very large dataset.

Kernel Methods

Kernels can be used to extend the above algorithms to non-parametric models (or models where the parameters form an infinite dimensional space). The corresponding procedure will no longer be truly online and instead involve storing all the data points, but is still faster than the brute force method. This discussion is restricted to the case of the square loss, though it can be extended to any convex loss. It can be shown by an easy induction that if X_i is the data matrix and w_i is the output after i steps of the SGD algorithm, then,

$$w_i = X_i^T c_i$$

where $c_i = ((c_i)_1, (c_i)_2, ..., (c_i)_i) \in \mathbb{R}^i$ and the sequence c_i satisfies the recursion:

$$c_0 = 0$$

$(c_i)_j = (c_{i-1})_j, j = 1, 2, ..., i-1$ and

$$(c_i)_i = \gamma_i \left(y_i - \sum_{j=1}^{i-1} (c_{i-1})_j \langle x_j, x_i \rangle \right)$$

Notice that here $\langle x_j, x_i \rangle$ is just the standard Kernel on $\mathbb{R}^d$, and the predictor is of the form

$$f_i(x) = \langle w_{i-1}, x \rangle = \sum_{j=1}^{i-1} (c_{i-1})_j \langle x_j, x \rangle.$$

Now, if a general kernel K is introduced instead and let the predictor be

$$f_i(x) = \sum_{j=1}^{i-1} (c_{i-1})_j K(x_j, x)$$

then the same proof will also show that predictor minimising the least squares loss is obtained by changing the above recursion to

$$(c_i)_i = \gamma_i \left(y_i - \sum_{j=1}^{i-1} (c_{i-1})_j K(x_j, x_i) \right)$$

The above expression requires storing all the data for updating c_i. The total time complexity for the recursion when evaluating for the n-th datapoint is $O(n^2 dk)$, where k is the cost of evaluating the kernel on a single pair of points. Thus, the use of the kernel has allowed the movement from a finite dimensional parameter space $w_i \in \mathbb{R}^d$ to a possibly infinite dimensional feature represented by a kernel K by instead performing the recursion on the space of parameters $c_i \in \mathbb{R}^i$, whose dimension is the same as the size of the training dataset. In general, this is a consequence of the representer theorem.

Progressive Learning

Progressive learning is an effective learning model which is demonstrated by the human learning process. It is the process of learning continuously from direct experience. Progressive learning technique (PLT) in machine learning can learn new classes/labels dynamically on the run. Though online learning can learn *new samples* of data that arrive sequentially, they cannot learn *new classes* of data being introduced to the model. The learning paradigm of progressive learning, is independent of the number of class constraints and it can learn new classes while still retaining the knowledge of previous classes. Whenever a new class (non-native to the knowledge learnt thus far) is encountered, the classifier gets remodeled automatically and the parameters are calculated in such a way that it retains the knowledge learnt thus far. This technique is suitable for real-world applications where the number of classes is often unknown and online learning from real-time data is required.

Online Convex Optimisation

In online convex optimisation (OCO), the hypothesis set and the loss functions are forced to be convex to obtain stronger learning bounds. The modified sequential game is now as follows:

For $t = 1, 2, ..., T$

- Learner receives input x_t

- Learner outputs w_t from a fixed convex set S

- Nature sends back a convex loss function $v_t : S \to \mathbb{R}$.

- Learner suffers loss $v_t(w_t)$ and updates its model

Thus, when regret is minimised, competition against the best weight vector $u \in H$ occurs. As an example, consider the case of online least squares linear regression. Here, the weight vectors come from the convex set $S = \mathbb{R}^d$, and nature sends back the convex loss function $v_t(w) = (\langle w, x_t \rangle - y_t)^2$. Note here that y_t is implicitly sent with v_t.

Some online prediction problems however cannot fit in the framework of OCO. For example, in online classification, the prediction domain and the loss functions are not convex. In such scenarios, two simple techniques for convexification are used: randomisation and surrogate loss functions.

Some simple online convex optimisation algorithms are:

Follow the Leader (FTL)

The simplest learning rule to try is to select (at the current step) the hypothesis that has the least loss over all past rounds. This algorithm is called Follow the leader, and is simply given round t by:

$$w_t = \mathrm{argmin}_{w \in S} \sum_{i=1}^{t-1} v_i(w)$$

This method can thus be looked as a greedy algorithm. For the case of online quadratic optimization (where the loss function is $v_t(w) = \| w - x_t \|_2^2$), one can show a regret bound that grows as $\log(T)$. However, similar bounds cannot be obtained for the FTL algorithm for other important families of models like online linear optimization. To do so, one modifies FTL by adding regularisation.

Follow the Regularised Leader (FTRL)

This is a natural modification of FTL that is used to stabilise the FTL solutions and obtain better regret bounds. A regularisation function $R : S \to \mathbb{R}$ is chosen and learning performed in round t as follows:

$$w_t = \mathrm{argmin}_{w \in S} \sum_{i=1}^{t-1} v_i(w) + R(w)$$

As a special example, consider the case of online linear optimisation i.e. where nature sends back loss functions of the form $v_t(w) = \langle w, z_t \rangle$. Also, let $S = \mathbb{R}^d$. Suppose the regularisation function $R(w) = \frac{1}{2\eta} \| w \|_2^2$ is chosen for some positive number η. Then, one can show that the regret minimising iteration becomes

$$w_{t+1} = -\eta \sum_{i=1}^{t} z_i = w_t - \eta z_t$$

Note that this can be rewritten as $w_{t+1} = w_t - \eta \nabla v_t(w_t)$, which looks exactly like online gradient descent.

If S is instead some convex subspace of $\mathbb{R}^d$, S would need to be projected onto, leading to the modified update rule

$$w_{t+1} = \Pi_S(-\eta \sum_{i=1}^{t} z_i) = \Pi_S(\eta \theta_{t+1})$$

This algorithm is known as lazy projection, as the vector θ_{t+1} accumulates the gradients. It is also known as Nesterov's dual averaging algorithm. In this scenario of linear loss functions and quadratic regularisation, the regret is bounded by $O(\sqrt{T})$, and thus the average regret goes to o as desired.

Online Subgradient Descent (OSD)

The above proved a regret bound for linear loss functions $v_t(w) = \langle w, z_t \rangle$. To generalise the algorithm to any convex loss function, the subgradient $\partial v_t(w_t)$ of v_t is used as a linear approximation to v_t near w_t, leading to the online subgradient descent algorithm:

Initialise parameter $\eta, w_1 = 0$

For $t = 1, 2, ..., T$

- Predict using w_t, receive f_t from nature.

- Choose $z_t \in \partial v_t(w_t)$

- If $S = \mathbb{R}^d$, update as $w_{t+1} = w_t - \eta z_t$

- If $S \subset \mathbb{R}^d$, project cumulative gradients onto S i.e. $w_{t+1} = \Pi_S(\eta \theta_{t+1}), \theta_{t+1} = \theta_t + z_t$

One can use the OSD algorithm to derive $O(\sqrt{T})$ regret bounds for the online version of SVM's for classification, which use the hinge loss $v_t(w) = \max\{0, 1 - y_t(w \cdot x_t)\}$

Other Algorithms

Quadratically regularised FTRL algorithms lead to lazily projected gradient algorithms as described above. To use the above for arbitrary convex functions and regularisers, one uses online mirror descent. Another algorithm is called prediction with expert advice. In this case, the hypothesis set consists of d functions. A distribution $w_t \in \Delta_d$ over the d experts is maintained, and prediction is performed by sampling an expert from this distribution. For the Euclidean regularisation, one can show a regret bound of $O(\sqrt{T})$, which can be improved further to a $O(\sqrt{\log T})$ bound by using a better regulariser.

Interpretations of Online Learning

The paradigm of online learning interestingly has different interpretations depending on the choice of the learning model, each of which has distinct implications about the predictive quality of the sequence of functions $f_1, f_2, \ldots, f_n$. The prototypical stochastic gradient descent algorithm is used for this discussion. As noted above, its recursion is given by

$$w_t = w_{t-1} - \gamma_t \nabla V(\langle w_{t-1}, x_t \rangle, y_t)$$

The first interpretation consider the stochastic gradient descent method as applied to the problem of minimizing the expected risk $I[w]$ defined above. Indeed, in the case of an infinite stream of data, since the examples $(x_1, y_1), (x_2, y_2), \ldots$ are assumed to be drawn i.i.d. from the distribution $p(x, y)$, the sequence of gradients of $V(\cdot, \cdot)$ in the above iteration are an i.i.d. sample of stochastic estimates of the gradient of the expected risk $I[w]$ and therefore one can apply complexity results for the stochastic gradient descent method to bound the deviation $I[w_t] - I[w^*]$, where w^* is the minimizer of $I[w]$. This interpretation is also valid in the case of a finite training set; although with multiple passes through the data the gradients are no longer independent, still complexity results can be obtained in special cases.

The second interpretation applies to the case of a finite training set and considers the SGD algorithm as an instance of incremental gradient descent method. In this case, one instead looks at the empirical risk:

$$I_n[w] = \frac{1}{n} \sum_{i=1}^{n} V(\langle w, x_i \rangle, y_i).$$

Since the gradients of $V(\cdot, \cdot)$ in the incremental gradient descent iterations are also stochastic estimates of the gradient of $I_n[w]$, this interpretation is also related to the stochastic gradient descent method, but applied to minimize the empirical risk as opposed to the expected risk. Since this interpretation concerns the empirical risk and not the expected risk, multiple passes through the data are readily allowed and actually lead to tighter bounds on the deviations $I_n[w_t] - I_n[w_n^*]$, where w_n^* is the minimizer of $I_n[w]$.

Implementations

- Vowpal Wabbit: Open-source fast out-of-core online learning system which is notable for supporting a number of machine learning reductions, importance weighting and a selection of different loss functions and optimisation algorithms. It uses the hashing trick for bounding the size of the set of features independent of the amount of training data.

- scikit-learn: Provides out-of-core implementations of algorithms for

  o Classification: Perceptron, SGD classifier, Naive bayes classifier.

  o Regression: SGD Regressor, Passive Aggressive regressor.

  o Clustering: Mini-batch k-means.

  o Feature extraction: Mini-batch dictionary learning, Incremental PCA.

Rule-based Machine Learning

Rule-based machine learning (RBML) is a term in computer science intended to encompass any machine learning method that identifies, learns, or evolves 'rules' to store, manipulate or apply. The defining characteristic of a rule-based machine learner is the identification and utilization of a set of relational rules that collectively represent the knowledge captured by the system. This is in contrast to other machine learners that commonly identify a singular model that can be universally applied to any instance in order to make a prediction.

Rule-based machine learning approaches include learning classifier systems, association rule learning, artificial immune systems, and any other method that relies on a set of rules, each covering contextual knowledge.

While rule-based machine learning is conceptually a type of rule-based system, it is distinct from traditional rule-based systems, which are often hand-crafted, and other rule-based decision makers. This is because rule-based machine learning applies some form of learning algorithm to automatically identify useful rules, rather than a human needing to apply prior domain knowledge to manually construct rules and curate a rule set.

Rules

Rules typically take the form of an {IF:THEN} expression, (e.g. {IF 'condition' THEN 'result'}, or as a more specific example, {IF 'red' AND 'octagon' THEN 'stop-sign'}). An individual rule is not in itself a model, since the rule is only applicable when its condition is satisfied. Therefore rule-based machine learning methods typically identify a set of rules that collectively comprise the prediction model, or the knowledge base.

Multiple Kernel Learning

Multiple kernel learning refers to a set of machine learning methods that use a predefined set of kernels and learn an optimal linear or non-linear combination of kernels as part of the algorithm. Reasons to use multiple kernel learning include a) the ability to select for an optimal kernel and parameters from a larger set of kernels, reducing bias due to kernel selection while allowing for more automated machine learning methods, and b) combining data from different sources (e.g. sound and images from a video) that have different notions of similarity and thus require different kernels. Instead of creating a new kernel, multiple kernel algorithms can be used to combine kernels already established for each individual data source.

Multiple kernel learning approaches have been used in many applications, such as event recognition in video, object recognition in images, and biomedical data fusion.

Algorithms

Multiple kernel learning algorithms have been developed for supervised, semi-supervised, as well as unsupervised learning. Most work has been done on the supervised learning case with linear combinations of kernels, however, many algorithms have been developed. The basic idea behind multiple kernel learning algorithms is to add an extra parameter to the minimization problem of the learning algorithm. As an example, consider the case of supervised learning of a linear combination of a set of n kernels $K' = \sum_{i=1}^{n} \beta_i K_i$. We introduce a new kernel $K' = \sum_{i=1}^{n} \beta_i K_i$, where β is a vector of coefficients for each kernel. Because the kernels are additive (due to properties of reproducing kernel Hilbert spaces), this new function is still a kernel. For a set of data X with labels Y, the minimization problem can then be written as

$$\min_{\beta,c} E(Y, K'c) + R(K, c)$$

where E is an error function and R is a regularization term. E is typically the square loss function (Tikhonov regularization) or the hinge loss function (for SVM algorithms), and R is usually an ℓ_n norm or some combination of the norms (i.e. elastic net regularization). This optimization problem can then be solved by standard optimization methods. Adaptations of existing techniques such as the Sequential Minimal Optimization have also been developed for multiple kernel SVM-based methods.

Supervised Learning

For supervised learning, there are many other algorithms that use different methods to learn the form of the kernel. The following categorization has been proposed by Gonen and Alpaydın (2011)

Fixed Rules Approaches

Fixed rules approaches such as the linear combination algorithm described above use rules to set the combination of the kernels. These do not require parameterization and use rules like summation and multiplication to combine the kernels. The weighting is learned in the algorithm. Other examples of fixed rules include pairwise kernels, which are of the form

$$k((x_{1i}, x_{1j}), (x_{2i}, x_{2j})) = k(x_{1i}, x_{2i})k(x_{1j}, x_{2j}) + k(x_{1i}, x_{2j})k(x_{1j}, x_{2i}).$$

These pairwise approaches have been used in predicting protein-protein interactions.

Heuristic Approaches

These algorithms use a combination function that is parameterized. The parameters are generally defined for each individual kernel based on single-kernel performance or some computation from the kernel matrix. Examples of these include the kernel from Tenabe et al. (2008). Letting π_m be the accuracy obtained using only K_m, and letting δ be a threshold less than the minimum of the single-kernel accuracies, we can define

$$\beta_m = \frac{\pi_m - \delta}{\sum_{h=1}^{n}(\pi_h - \delta)}$$

Other approaches use a definition of kernel similarity, such as

$$A(K_1,K_2) = \frac{<K_1,K_2>}{\sqrt{<K_1,K_1><K_2,K_2>}}$$

Using this measure, Qui and Lane (2009) used the following heuristic to define

$$\beta_m = \frac{A(K_m,YY^T)}{\sum_{h=1}^{n} A(K_h,YY^T)}$$

Optimization Approaches

These approaches solve an optimization problem to determine parameters for the kernel combination function. This has been done with similarity measures and structural risk minimization approaches. For similarity measures such as the one defined above, the problem can be formulated as follows:

$$\max_{\beta,tr(K'_{tra})=1,K'\geq 0} A(K'_{tra},YY^T).$$

where K'_{tra} is the kernel of the training set.

Structural risk minimization approaches that have been used include linear approaches, such as that used by Lanckriet et al. (2002). We can define the implausibility of a kernel $\omega(K)$ to be the value of the objective function after solving a canonical SVM problem. We can then solve the following minimization problem:

$$\min_{tr(K'_{tra})=c} \omega(K'_{tra})$$

where   is a positive constant. Many other variations exist on the same idea, with different methods of refining and solving the problem, e.g. with nonnegative weights for individual kernels and using non-linear combinations of kernels.

Bayesian Approaches

Bayesian approaches put priors on the kernel parameters and learn the parameter values from the priors and the base algorithm. For example, the decision function can be written as

$$f(x) = \sum_{i=0}^{n} \alpha_i \sum_{m=1}^{p} \eta_m K_m(x_i^m,x^m)$$

η can be modeled with a Dirichlet prior and α can be modeled with a zero-mean Gaussian and an inverse gamma variance prior. This model is then optimized using a customized multinomial probit approach with a Gibbs sampler.

These methods have been used successfully in applications such as protein fold recognition and protein homology problems

Boosting Approaches

Boosting approaches add new kernels iteratively until some stopping criteria that is a function of performance is reached. An example of this is the MARK model developed by Bennett et al. (2002)

$$f(x) = \sum_{i=1}^{N}\sum_{m=1}^{P} \alpha_i^m K_m(x_i^m, x^m) + b$$

The parameters α_i^m and b are learned by gradient descent on a coordinate basis. In this way, each iteration of the descent algorithm identifies the best kernel column to choose at each particular iteration and adds that to the combined kernel. The model is then rerun to generate the optimal weights α_i and b.

Semisupervised Learning

Semisupervised learning approaches to multiple kernel learning are similar to other extensions of supervised learning approaches. An inductive procedure has been developed that uses a log-likelihood empirical loss and group LASSO regularization with conditional expectation consensus on unlabeled data for image categorization. We can define the problem as follows. Let $L = (x_i, y_i)$ be the labeled data, and let $U = x_i$ be the set of unlabeled data. Then, we can write the decision function as follows.

$$f(x) = \alpha_0 + \sum_{i=1}^{|L|} \alpha_i K_i(x)$$

The problem can be written as

$$\min_f L(f) + \lambda R(f) + \gamma \Theta(f)$$

where L is the loss function (weighted negative log-likelihood in this case), R is the regularization parameter (Group LASSO in this case), and Θ is the conditional expectation consensus (CEC) penalty on unlabeled data. The CEC penalty is defined as follows. Let the marginal kernel density for all the data be

$$g_m^\pi(x) = <\phi_m^\pi, \psi_m(x)>$$

where $\psi_m(x) = [K_m(x_1, x), \ldots, K_m(x_L, x)]^T$ (the kernel distance between the labeled data and all of the labeled and unlabeled data) and ϕ_m^π is a non-negative random vector with a 2-norm of 1. The value of Π is the number of times each kernel is projected. Expectation regularization is then performed on the MKD, resulting in a reference expectation $q_m^{pi}(y \mid g_m^\pi(x))$ and model expectation $p_m^\pi(f(x) \mid g_m^\pi(x))$. Then, we define

$$\Theta = \frac{1}{\Pi}\sum_{\pi=1}^{\Pi}\sum_{m=1}^{M} D(q_m^{pi}(y \mid g_m^\pi(x)) \| p_m^\pi(f(x) \mid g_m^\pi(x)))$$

where $D(Q \| P) = \sum_i Q(i) \ln \frac{Q(i)}{P(i)}$ is the Kullback-Leibler divergence. The combined minimization problem is optimized using a modified block gradient descent algorithm.

Unsupervised Learning

Unsupervised multiple kernel learning algorithms have also been proposed by Zhuang et al. The problem is defined as follows. Let $U = x_i$ be a set of unlabeled data. The kernel definition is the linear combined kernel $K' = \sum_{i=1}^{M} \beta_i K_m$. In this problem, the data needs to be "clustered" into groups based on the kernel distances. Let B_i be a group or cluster of which x_i is a member. We define the loss function as $\sum_{i=1}^{n} \left\| x_i - \sum_{x_j \in B_i} K(x_i, x_j) x_j \right\|^2$. Furthermore, we minimize the distortion by minimizing $\sum_{i=1}^{n} \sum_{x_j \in B_i} K(x_i, x_j) \left\| x_i - x_j \right\|^2$. Finally, we add a regularization term to avoid overfitting. Combining these terms, we can write the minimization problem as follows.

$$\min_{\beta, B} \sum_{i=1}^{n} \left\| x_i - \sum_{x_j \in B_i} K(x_i, x_j) x_j \right\|^2 + \gamma_1 \sum_{i=1}^{n} \sum_{x_j \in B_i} K(x_i, x_j) \left\| x_i - x_j \right\|^2 + \gamma_2 \sum_i |B_i|$$

where. One formulation of this is defined as follows. Let $D \in {0,1}^{n \times n}$ be a matrix such that $D_{ij} = 1$ means that x_i and x_j are neighbors. Then, $B_i = x_j : D_{ij} = 1$. Note that these groups must be learned as well. Zhuang et al. solve this problem by an alternating minimization method for K and the groups B_i.

MKL Libraries

Available MKL libraries include

- SPG-GMKL: A scalable C++ MKL SVM library that can handle a million kernels.

- GMKL: Generalized Multiple Kernel Learning code in MATLAB, does ℓ_1 and ℓ_2 regularization for supervised learning.

- (Another) GMKL: A different MATLAB MKL code that can also perform elastic net regularization

- SMO-MKL: C++ source code for a Sequential Minimal Optimization MKL algorithm. Does $p-n$-n orm regularization.

- SimpleMKL: A MATLAB code based on the SimpleMKL algorithm for MKL SVM.

Temporal Difference Learning

Temporal difference (TD) learning is a prediction-based machine learning method. It has primarily been used for the reinforcement learning problem, and is said to be "a combination of Monte Carlo ideas and dynamic programming (DP) ideas." TD resembles a Monte Carlo method because it learns by sampling the environment according to some *policy*, and is related to dynamic programming techniques as it approximates its current estimate based on previously learned estimates (a process known as bootstrapping). The TD learning algorithm is related to the temporal difference model of animal learning.

As a prediction method, TD learning considers that subsequent predictions are often correlated in some sense. In standard supervised predictive learning, one learns only from actually observed values: A prediction is made, and when the observation is available, the prediction mechanism is adjusted to better match the observation. As elucidated by Richard Sutton, the core idea of TD learning is that one adjusts predictions to match other, more accurate, predictions about the future. This procedure is a form of bootstrapping, as illustrated with the following example:

> "Suppose you wish to predict the weather for Saturday, and you have some model that predicts Saturday's weather, given the weather of each day in the week. In the standard case, you would wait until Saturday and then adjust all your models. However, when it is, for example, Friday, you should have a pretty good idea of what the weather would be on Saturday - and thus be able to change, say, Monday's model before Saturday arrives."

Mathematically speaking, both in a standard and a TD approach, one would try to optimize some cost function, related to the error in our predictions of the expectation of some random variable, E[z]. However, while in the standard approach one in some sense assumes E[z] = z (the actual observed value), in the TD approach we use a model. For the particular case of reinforcement learning, which is the major application of TD methods, z is the total return and E[z] is given by the Bellman equation of the return.

Mathematical Formulation

Let r_t be the reward (return) on time step t. Let $\overline{V}_t$ be the correct prediction that is equal to the discounted sum of all future reward. The discounting is done by powers of factor of γ such that reward at distant time step is less important.

$$\overline{V}_t = \sum_{i=0}^{\infty} \gamma^i r_{t+i}$$

where $0 \leq \gamma < 1$. This formula can be expanded

$$\overline{V}_t = r_t + \sum_{i=1}^{\infty} \gamma^i r_{t+i}$$

by changing the index of i to start from 0.

$$\overline{V}_t = r_t + \sum_{i=0}^{\infty} \gamma^{i+1} r_{t+i+1}$$

$$\overline{V}_t = r_t + \gamma \sum_{i=0}^{\infty} \gamma^i r_{t+i+1}$$

$$\overline{V}_t = r_t + \gamma \overline{V}_{t+1}$$

Thus, the reward is the difference between the correct prediction and the current prediction.

$$r_t = \overline{V}_t - \gamma \overline{V}_{t+1}$$

TD-Lambda

TD-Lambda is a learning algorithm invented by Richard S. Sutton based on earlier work on temporal difference learning by Arthur Samuel. This algorithm was famously applied by Gerald Tesauro to create TD-Gammon, a program that learned to play the game of backgammon at the level of expert human players.

The lambda (λ) parameter refers to the trace decay parameter, with $0 \leq \lambda \leq 1$. Higher settings lead to longer lasting traces; that is, a larger proportion of credit from a reward can be given to more distant states and actions when λ is higher, with $\lambda = 1$ producing parallel learning to Monte Carlo RL algorithms.

TD Algorithm in Neuroscience

The TD algorithm has also received attention in the field of neuroscience. Researchers discovered that the firing rate of dopamine neurons in the ventral tegmental area (VTA) and substantia nigra (SNc) appear to mimic the error function in the algorithm. The error function reports back the difference between the estimated reward at any given state or time step and the actual reward received. The larger the error function, the larger the difference between the expected and actual reward. When this is paired with a stimulus that accurately reflects a future reward, the error can be used to associate the stimulus with the future reward.

Dopamine cells appear to behave in a similar manner. In one experiment measurements of dopamine cells were made while training a monkey to associate a stimulus with the reward of juice. Initially the dopamine cells increased firing rates when the monkey received juice, indicating a difference in expected and actual rewards. Over time this increase in firing back propagated to the earliest reliable stimulus for the reward. Once the monkey was fully trained, there was no increase in firing rate upon presentation of the predicted reward. Continually, the firing rate for the dopamine cells decreased below normal activation when the expected reward was not produced. This mimics closely how the error function in TD is used for reinforcement learning.

The relationship between the model and potential neurological function has produced research attempting to use TD to explain many aspects of behavioral research. It has also been used to study conditions such as schizophrenia or the consequences of pharmacological manipulations of dopamine on learning.

Association Rule Learning

Association rule learning is a rule-based machine learning method for discovering interesting relations between variables in large databases. It is intended to identify strong rules discovered in databases using some measures of interestingness. Based on the concept of strong rules, Rakesh Agrawal, Tomasz Imieliński and Arun Swami introduced association rules for discovering regularities between products in large-scale transaction data recorded by point-of-sale (POS) systems in supermarkets. For example, the rule {onions, potatoes} ⇒ {burger} found in the sales data of a supermarket would indicate that if a customer buys onions and potatoes together, they are likely to also buy hamburger meat. Such information can be used as the basis for decisions about marketing activities such as, e.g., promotional pricing or product placements. In addition to the above example from market basket analysis association rules are employed today in many application areas including Web usage mining, intrusion detection, continuous production, and bioinformatics. In contrast with sequence mining, association rule learning typically does not consider the order of items either within a transaction or across transactions.

Definition

Example database with 5 transactions and 5 items					
Transaction Id	Milk	Bread	Butter	Beer	Diapers
1	1	1	0	0	0
2	0	0	1	0	0
3	0	0	0	1	1
4	1	1	1	0	0
5	0	1	0	0	0

Following the original definition by Agrawal, Imieliński, Swami the problem of association rule mining is defined as:

Let $I = \{i_1, i_2, \ldots, i_n\}$ be a set of n binary attributes called *items*.

Let $D = \{t_1, t_2, \ldots, t_m\}$ be a set of transactions called the *database*.

Each *transaction* in D has a unique transaction ID and contains a subset of the items in I.

A *rule* is defined as an implication of the form:

$X \Rightarrow Y$, where $X, Y \subseteq I$.

In Agrawal, Imieliński, Swami a *rule* is defined only between a set and a single item, $X \Rightarrow i_j$ for $i_j \in I$.

Every rule is composed by two different sets of items, also known as *itemsets*, X and Y, where X is called *antecedent* or left-hand-side (LHS) and Y *consequent* or right-hand-side (RHS).

To illustrate the concepts, we use a small example from the supermarket domain. The set of items is $I = \{milk, bread, butter, beer, diapers\}$ and in the table is shown a small database containing the items, where, in each entry, the value 1 means the presence of the item in the corresponding transaction, and the value 0 represents the absence of an item in that transaction.

An example rule for the supermarket could be {butter, bread} $\Rightarrow$ {milk} meaning that if butter and bread are bought, customers also buy milk.

Note: this example is extremely small. In practical applications, a rule needs a support of several hundred transactions before it can be considered statistically significant, and datasets often contain thousands or millions of transactions.

Useful Concepts

In order to select interesting rules from the set of all possible rules, constraints on various measures of significance and interest are used. The best-known constraints are minimum thresholds on support and confidence.

Let X be an itemset, $X \Rightarrow Y$ an association rule and T a set of transactions of a given database.

Support

Support is an indication of how frequently the itemset appears in the dataset.

The support of X with respect to T is defined as the proportion of transactions t in the dataset which contains the itemset X.

$$\text{supp}(X) = \frac{|\{t \in T; X \subseteq t\}|}{|T|}$$

In the example dataset, the itemset $X = \{\text{beer}, \text{diapers}\}$ h as a support of $1/5 = 0.2$ since it occurs in 20% of all transactions (1 out of 5 transactions). The argument of supp() is a set of preconditions, and thus becomes more restrictive as it grows (instead of more inclusive).

Confidence

Confidence is an indication of how often the rule has been found to be true.

The *confidence* value of a rule, $X \Rightarrow Y$, with respect to a set of transactions T, is the proportion of the transactions that contains X which also contains Y.

Confidence is defined as:

$$\text{conf}(X \Rightarrow Y) = \text{supp}(X \cup Y) / \text{supp}(X).$$

For example, the rule {butter, bread} $\Rightarrow$ {milk} has a confidence of $0.2/0.2 = 1.0$ in the database, which means that for 100% of the transactions containing butter and bread the rule is correct (100% of the times a customer buys butter and bread, milk is bought as well).

Note that supp($X \cup Y$) means the support of the union of the items in X and Y. This is somewhat confusing since we normally think in terms of probabilities of events and not sets of items. We can rewrite supp($X \cup Y$) as the probability $P(E_X \wedge E_Y)$, where E_X and E_Y are the events that a transaction contains itemset X and Y, respectively.

Thus confidence can be interpreted as an estimate of the conditional probability $P(E_Y | E_X)$, the probability of finding the RHS of the rule in transactions under the condition that these transactions also contain the LHS.

Lift

The *lift* of a rule is defined as:

$$\text{lift}(X \Rightarrow Y) = \frac{\text{supp}(X \cup Y)}{\text{supp}(X) \times \text{supp}(Y)}$$

or the ratio of the observed support to that expected if X and Y were independent.

For example, the rule $\{\text{milk}, \text{bread}\} \Rightarrow \{\text{butter}\}$ has a lift of $\dfrac{0.2}{0.4 \times 0.4} = 1.25$.

If the rule had a lift of 1, it would imply that the probability of occurrence of the antecedent and that of the consequent are independent of each other. When two events are independent of each other, no rule can be drawn involving those two events.

If the lift is > 1, that lets us know the degree to which those two occurrences are dependent on one another, and makes those rules potentially useful for predicting the consequent in future data sets.

The value of lift is that it considers both the confidence of the rule and the overall data set.

Conviction

The *conviction* of a rule is defined as $\text{conv}(X \Rightarrow Y) = \dfrac{1 - \text{supp}(Y)}{1 - \text{conf}(X \Rightarrow Y)}.$

For example, the rule $\{\text{milk}, \text{bread}\} \Rightarrow \{\text{butter}\}$ has a conviction of $\dfrac{1 - 0.4}{1 - 0.5} = 1.2$, and can be interpreted as the ratio of the expected frequency that X occurs without Y (that is to say, the frequency that the rule makes an incorrect prediction) if X and Y were independent divided by the observed frequency of incorrect predictions. In this example, the conviction value of 1.2 shows that the rule $\{\text{milk}, \text{bread}\} \Rightarrow \{\text{butter}\}$ would be correct 20% more often (1.2 times as often) if the association between X and Y was purely random chance.

Process

Association rules are usually required to satisfy a user-specified minimum support and a user-specified minimum confidence at the same time. Association rule generation is usually split up into two separate steps:

1. A minimum support threshold is applied to find all *frequent itemsets* in a database.

2. A minimum confidence constraint is applied to these frequent itemsets in order to form rules.

While the second step is straightforward, the first step needs more attention.

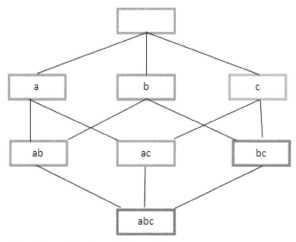

Frequent itemset lattice, where the color of the box indicates how many transactions contain the combination of items. Note that lower levels of the lattice can contain at most the minimum number of their parents' items; e.g. {ac} can have only at most $\min(a,c)$ items. This is called the *downward-closure property*.

Finding all frequent itemsets in a database is difficult since it involves searching all possible itemsets (item combinations). The set of possible itemsets is the power set over I and has size $2^n - 1$ (excluding the empty set which is not a valid itemset). Although the size of the power-set grows exponentially in the number of items n in I, efficient search is possible using the *downward-closure property* of support (also called *anti-monotonicity*) which guarantees that for a frequent itemset, all its subsets are also frequent and thus no infrequent itemset can be a subset of a frequent itemset. Exploiting this property, efficient algorithms (e.g., Apriori and Eclat) can find all frequent itemsets.

History

The concept of association rules was popularised particularly due to the 1993 article of Agrawal et al., which has acquired more than 18,000 citations according to Google Scholar, as of August 2015, and is thus one of the most cited papers in the Data Mining field. However, it is possible that what is now called "association rules" is similar to what appears in the 1966 paper on GUHA, a general data mining method developed by Petr Hájek et al.

An early (circa 1989) use of minimum support and confidence to find all association rules is the Feature Based Modeling framework, which found all rules with $\text{supp}(X)$ and $\text{conf}(X \Rightarrow Y)$ greater than user defined constraints.

Alternative Measures of Interestingness

In addition to confidence, other measures of *interestingness* for rules have been proposed. Some popular measures are:

- All-confidence

- Collective strength

- Conviction

- Leverage

- Lift (originally called interest)

Several more measures are presented and compared by Tan et al. and by Hahsler. Looking for techniques that can model what the user has known (and using these models as interestingness measures) is currently an active research trend under the name of "Subjective Interestingness."

Statistically Sound Associations

One limitation of the standard approach to discovering associations is that by searching massive numbers of possible associations to look for collections of items that appear to be associated, there is a large risk of finding many spurious associations. These are collections of items that co-occur with unexpected frequency in the data, but only do so by chance. For example, suppose we are considering a collection of 10,000 items and looking for rules containing two items in the left-hand-side and 1 item in the right-hand-side. There are approximately 1,000,000,000,000 such rules. If we apply a statistical test for independence with a significance level of 0.05 it means there is only a 5% chance of accepting a rule if there is no association. If we assume there are no associations, we should nonetheless expect to find 50,000,000,000 rules. Statistically sound association discovery controls this risk, in most cases reducing the risk of finding *any* spurious associations to a user-specified significance level.

Algorithms

Many algorithms for generating association rules have been proposed.

Some well-known algorithms are Apriori, Eclat and FP-Growth, but they only do half the job, since they are algorithms for mining frequent itemsets. Another step needs to be done after to generate rules from frequent itemsets found in a database.

Apriori Algorithm

Apriori uses a breadth-first search strategy to count the support of itemsets and uses a candidate generation function which exploits the downward closure property of support.

Eclat Algorithm

Eclat (alt. ECLAT, stands for Equivalence Class Transformation) is a depth-first search algorithm using set intersection. It is a naturally elegant algorithm suitable for both sequential as well as parallel execution with locality-enhancing properties. It was first introduced by Zaki, Parthasarathy, Li and Ogihara in a series of papers written in 1997.

FP-growth Algorithm

FP stands for frequent pattern.

In the first pass, the algorithm counts occurrence of items (attribute-value pairs) in the dataset, and stores them to 'header table'. In the second pass, it builds the FP-tree structure by inserting instances. Items in each instance have to be sorted by descending order of their frequency in the dataset, so that the tree can be processed quickly. Items in each instance that do not meet minimum coverage threshold are discarded. If many instances share most frequent items, FP-tree provides high compression close to tree root.

Recursive processing of this compressed version of main dataset grows large item sets directly, instead of generating candidate items and testing them against the entire database. Growth starts from the bottom of the header table (having longest branches), by finding all instances matching given condition. New tree is created, with counts projected from the original tree corresponding to the set of instances that are conditional on the attribute, with each node getting sum of its children counts. Recursive growth ends when no individual items conditional on the attribute meet minimum support threshold, and processing continues on the remaining header items of the original FP-tree.

Once the recursive process has completed, all large item sets with minimum coverage have been found, and association rule creation begins.

Others

AprioriDP

AprioriDP utilizes Dynamic Programming in Frequent itemset mining. The working principle is to eliminate the candidate generation like FP-tree, but it stores support count in specialized data structure instead of tree.

Context Based Association Rule Mining Algorithm

CBPNARM is an algorithm, developed in 2013, to mine association rules on the basis of context. It uses context variable on the basis of which the support of an itemset is changed on the basis of which the rules are finally populated to the rule set.

Node-set-based Algorithms

FIN, PrePost and PPV are three algorithms based on node sets. They use nodes in a coding FP-tree to represent itemsets, and employ a depth-first search strategy to discovery frequent itemsets using "intersection" of node sets.

GUHa Procedure ASSOC

GUHA is a general method for exploratory data analysis that has theoretical foundations in observational calculi.

The ASSOC procedure is a GUHA method which mines for generalized association rules using fast bitstrings operations. The association rules mined by this method are more general than those output by apriori, for example "items" can be connected both with conjunction and disjunctions and the relation between antecedent and consequent of the rule is not restricted to setting minimum support and confidence as in apriori: an arbitrary combination of supported interest measures can be used.

OPUS Search

OPUS is an efficient algorithm for rule discovery that, in contrast to most alternatives, does not require either monotone or anti-monotone constraints such as minimum support. Initially used to find rules for a fixed consequent it has subsequently been extended to find rules with any item as a consequent. OPUS search is the core technology in the popular Magnum Opus association discovery system.

Lore

A famous story about association rule mining is the "beer and diaper" story. A purported survey of behavior of supermarket shoppers discovered that customers (presumably young men) who buy diapers tend also to buy beer. This anecdote became popular as an example of how unexpected association rules might be found from everyday data. There are varying opinions as to how much of the story is true. Daniel Powers says:

In 1992, Thomas Blischok, manager of a retail consulting group at Teradata, and his staff prepared an analysis of 1.2 million market baskets from about 25 Osco Drug stores. Database queries were developed to identify affinities. The analysis "did discover that between 5:00 and 7:00 p.m. that consumers bought beer and diapers". Osco managers did NOT exploit the beer and diapers relationship by moving the products closer together on the shelves.

Other Types of Association Rule Mining

Multi-Relation Association Rules: Multi-Relation Association Rules (MRAR) are association rules where each item may have several relations. These relations indicate indirect relationship between the entities. Consider the following MRAR where the first item consists of three relations *live in*, *nearby* and *humid*: "Those who *live in* a place which is *near by* a city with *humid* climate type and also are *younger* than 20 -> their *health condition* is good". Such association rules are extractable from RDBMS data or semantic web data.

Context Based Association Rules are a form of association rule. Context Based Association Rules claims more accuracy in association rule mining by considering a hidden variable named context variable which changes the final set of association rules depending upon the value of context variables. For example the baskets orientation in market basket analysis reflects an odd pattern in the early days of month. This might be because of abnormal context i.e. salary is drawn at the start of the month

Contrast set learning is a form of associative learning. Contrast set learners use rules that differ meaningfully in their distribution across subsets.

Weighted class learning is another form of associative learning in which weight may be assigned to classes to give focus to a particular issue of concern for the consumer of the data mining results.

High-order pattern discovery facilitate the capture of high-order (polythetic) patterns or event associations that are intrinsic to complex real-world data.

K-optimal pattern discovery provides an alternative to the standard approach to association rule learning that requires that each pattern appear frequently in the data.

Approximate Frequent Itemset mining is a relaxed version of Frequent Itemset mining that allows some of the items in some of the rows to be 0.

Generalized Association Rules hierarchical taxonomy (concept hierarchy)

Quantitative Association Rules categorical and quantitative data

Interval Data Association Rules e.g. partition the age into 5-year-increment ranged

Sequential pattern mining discovers subsequences that are common to more than minsup sequences in a sequence database, where minsup is set by the user. A sequence is an ordered list of transactions.

Subspace Clustering, a specific type of Clustering high-dimensional data, is in many variants also based on the downward-closure property for specific clustering models.

Warmr is shipped as part of the ACE data mining suite. It allows association rule learning for first order relational rules.

References

- Zimek, Arthur; Assent, Ira; Vreeken, Jilles (2014). "Frequent Pattern Mining Algorithms for Data Clustering": 403–423. doi:10.1007/978-3-319-07821-2_16

- M., Weiss, S.; N., Indurkhya, (1995-01-01). "Rule-based Machine Learning Methods for Functional Prediction". Journal of Artificial Intelligence Research. 3. doi:10.1613/jair.199 (inactive 2017-01-20)

- Yin, Harold J. Kushner, G. George (2003). Stochastic approximation and recursive algorithms and applications (Second edition. ed.). New York: Springer. pp. 8–12. ISBN 978-0-387-21769-7

- Han (2000). "Mining Frequent Patterns Without Candidate Generation". Proceedings of the 2000 ACM SIGMOD International Conference on Management of Data. SIGMOD '00: 1–12. doi:10.1145/342009.335372

- Urbanowicz, Ryan J.; Moore, Jason H. (2009-09-22). "Learning Classifier Systems: A Complete Introduction, Review, and Roadmap". Journal of Artificial Evolution and Applications. 2009: 1–25. ISSN 1687-6229. doi:10.1155/2009/736398

- Bottou, Léon (1998). "Online Algorithms and Stochastic Approximations". Online Learning and Neural Networks. Cambridge University Press. ISBN 978-0-521-65263-6

- Zaki, M. J. (2000). "Scalable algorithms for association mining". IEEE Transactions on Knowledge and Data Engineering. 12 (3): 372–390. doi:10.1109/69.846291

- Gert R. G. Lanckriet, Nello Cristianini, Peter Bartlett, Laurent El Ghaoui, and Michael I. Jordan. Learning the kernel matrix with semidefinite programming. Journal of Machine Learning Research, 5:27–72, 2004a

- Tan, Pang-Ning; Michael, Steinbach; Kumar, Vipin (2005). "Chapter 6. Association Analysis: Basic Concepts and Algorithms" (PDF). Introduction to Data Mining. Addison-Wesley. ISBN 0-321-32136-7

- Hipp, J.; Güntzer, U.; Nakhaeizadeh, G. (2000). "Algorithms for association rule mining --- a general survey and comparison". ACM SIGKDD Explorations Newsletter. 2: 58. doi:10.1145/360402.360421

- Alain Rakotomamonjy, Francis Bach, Stephane Canu, Yves Grandvalet. SimpleMKL. Journal of Machine Learning Research, Microtome Publishing, 2008, 9, pp.2491-2521

- Hájek, Petr; Feglar, Tomas; Rauch, Jan; and Coufal, David; The GUHA method, data preprocessing and mining, Database Support for Data Mining Applications, Springer, 2004, ISBN 978-3-540-22479-2

- Hahsler, Michael (2005). "Introduction to arules – A computational environment for mining association rules and frequent item sets" (PDF)

- Hájek, Petr; Havránek, Tomáš (1978). Mechanizing Hypothesis Formation: Mathematical Foundations for a General Theory. Springer-Verlag. ISBN 3-540-08738-9

- Z. H. Deng and Z. Wang. A New Fast Vertical Method for Mining Frequent Patterns [4]. International Journal of Computational Intelligence Systems, 3(6): 733 - 744, 2010

- Venkatesan, Rajasekar; Meng Joo, Er (2016). "A novel progressive learning technique for multi-class classification". Neurocomputing. 207: 310–321. doi:10.1016/j.neucom.2016.05.006

- Zaki, Mohammed J. (2001); SPADE: An Efficient Algorithm for Mining Frequent Sequences, Machine Learning Journal, 42, pp. 31–60

Machine Learning Algorithms

Important machine learning algorithms include expectation-maximization algorithm, structured kNN, wake-sleep algorithm, etc. Structured kNN algorithms generalizes the kNN classifier. The topics discussed in the chapter are of great importance to broaden the existing knowledge on machine learning.

Expectation–Maximization Algorithm

In statistics, an expectation–maximization (EM) algorithm is an iterative method to find maximum likelihood or maximum a posteriori (MAP) estimates of parameters in statistical models, where the model depends on unobserved latent variables. The EM iteration alternates between performing an expectation (E) step, which creates a function for the expectation of the log-likelihood evaluated using the current estimate for the parameters, and a maximization (M) step, which computes parameters maximizing the expected log-likelihood found on the E step. These parameter-estimates are then used to determine the distribution of the latent variables in the next E step.

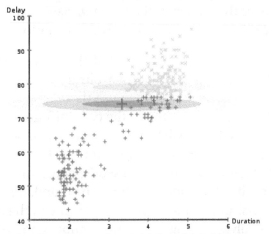

EM clustering of Old Faithful eruption data. The random initial model (which, due to the different scales of the axes, appears to be two very flat and wide spheres) is fit to the observed data. In the first iterations, the model changes substantially, but then converges to the two modes of the geyser.

History

The EM algorithm was explained and given its name in a classic 1977 paper by Arthur Dempster, Nan Laird, and Donald Rubin. They pointed out that the method had been "proposed many times in special circumstances" by earlier authors. A very detailed treatment of the EM method for exponential families was published by Rolf Sundberg in his thesis and several papers following his collaboration with Per Martin-Löf and Anders Martin-Löf. The Dempster-Laird-Rubin paper in

1977 generalized the method and sketched a convergence analysis for a wider class of problems. Regardless of earlier inventions, the innovative Dempster-Laird-Rubin paper in the *Journal of the Royal Statistical Society* received an enthusiastic discussion at the Royal Statistical Society meeting with Sundberg calling the paper "brilliant". The Dempster-Laird-Rubin paper established the EM method as an important tool of statistical analysis.

The convergence analysis of the Dempster-Laird-Rubin paper was flawed and a correct convergence analysis was published by C.F. Jeff Wu in 1983. Wu's proof established the EM method's convergence outside of the exponential family, as claimed by Dempster-Laird-Rubin.

Introduction

The EM algorithm is used to find (locally) maximum likelihood parameters of a statistical model in cases where the equations cannot be solved directly. Typically these models involve latent variables in addition to unknown parameters and known data observations. That is, either missing values exist among the data, or the model can be formulated more simply by assuming the existence of further unobserved data points. For example, a mixture model can be described more simply by assuming that each observed data point has a corresponding unobserved data point, or latent variable, specifying the mixture component to which each data point belongs.

Finding a maximum likelihood solution typically requires taking the derivatives of the likelihood function with respect to all the unknown values, the parameters and the latent variables, and simultaneously solving the resulting equations. In statistical models with latent variables, this is usually impossible. Instead, the result is typically a set of interlocking equations in which the solution to the parameters requires the values of the latent variables and vice versa, but substituting one set of equations into the other produces an unsolvable equation.

The EM algorithm proceeds from the observation that the following is a way to solve these two sets of equations numerically. One can simply pick arbitrary values for one of the two sets of unknowns, use them to estimate the second set, then use these new values to find a better estimate of the first set, and then keep alternating between the two until the resulting values both converge to fixed points. It's not obvious that this will work at all, but it can be proven that in this context it does, and that the derivative of the likelihood is (arbitrarily close to) zero at that point, which in turn means that the point is either a maximum or a saddle point. In general, multiple maxima may occur, with no guarantee that the global maximum will be found. Some likelihoods also have singularities in them, i.e., nonsensical maxima. For example, one of the *solutions* that may be found by EM in a mixture model involves setting one of the components to have zero variance and the mean parameter for the same component to be equal to one of the data points.

Description

Given the statistical model which generates a set $\mathbf{X}$ of observed data, a set of unobserved latent data or missing values $\mathbf{Z}$, and a vector of unknown parameters θ, along with a likelihood function $L(\theta; X, Z) = p(X, Z \mid \theta)$, the maximum likelihood estimate (MLE) of the unknown parameters is determined by the marginal likelihood of the observed data.

$$L(\theta; X) = p(X \mid \theta) = \int p(X, Z \mid \theta) dZ$$

However, this quantity is often intractable (e.g. if $\mathbf{Z}$ is a sequence of events, so that the number of values grows exponentially with the sequence length, making the exact calculation of the sum extremely difficult).

The EM algorithm seeks to find the MLE of the marginal likelihood by iteratively applying these two steps:

> *Expectation step (E step)*: Calculate the expected value of the log likelihood function, with respect to the conditional distribution of $\mathbf{Z}$ given $\mathbf{X}$ under the current estimate of the parameters $\theta^{(t)}$:

$$Q(\theta \,|\, \theta^{(t)}) = E_{\mathbf{Z}|\mathbf{X},\theta^{(t)}}\left[\log L(\theta; \mathbf{X}, \mathbf{Z})\right]$$

> *Maximization step (M step)*: Find the parameter that maximizes this quantity:

$$\theta^{(t+1)} = \underset{\theta}{\arg\max}\ Q(\theta \,|\, \theta^{(t)})$$

In typical models to which EM is applied:

1. The observed data points $\mathbf{X}$ may be discrete (taking values in a finite or countably infinite set) or continuous (taking values in an uncountably infinite set). Associated with each data point may be a vector of observations.

2. The missing values (aka latent variables) $\mathbf{Z}$ are discrete, drawn from a fixed number of values, and with one latent variable per observed data point.

3. The parameters are continuous, and are of two kinds: Parameters that are associated with all data points, and those associated with a specific value of a latent variable (i.e., associated with all data points which corresponding latent variable has that value).

However, it is possible to apply EM to other sorts of models.

The motive is as follows. If the value of the parameters θ is known, usually the value of the latent variables $\mathbf{Z}$ can be found by maximizing the log-likelihood over all possible values of $\mathbf{Z}$, either simply by iterating over $\mathbf{Z}$ or through an algorithm such as the Viterbi algorithm for hidden Markov models. Conversely, if we know the value of the latent variables $\mathbf{Z}$, we can find an estimate of the parameters θ fairly easily, typically by simply grouping the observed data points according to the value of the associated latent variable and averaging the values, or some function of the values, of the points in each group. This suggests an iterative algorithm, in the case where both θ and $\mathbf{Z}$ are unknown:

1. First, initialize the parameters θ to some random values.

2. Compute the best value for   given these parameter values.

3. Then, use the just-computed values of $\mathbf{Z}$ to compute a better estimate for the parameters θ. Parameters associated with a specific value of $\mathbf{Z}$ will use only those data points which associated latent variable has that value.

4. Iterate steps 2 and 3 until convergence.

The algorithm as just described monotonically approaches a local minimum of the cost function, and is commonly called *hard EM*. The k-means algorithm is an example of this class of algorithms.

However, somewhat better methods exist. Rather than making a hard choice for $\mathbf{Z}$ given the current parameter values and averaging only over the set of data points associated with some value of   , instead, determine the probability of each possible value of $\mathbf{Z}$ for each data point, and then use the probabilities associated with some value of $\mathbf{Z}$ to compute a weighted average over the whole set of data points. The resulting algorithm is commonly called *soft EM*, and is the type of algorithm normally associated with EM. The counts used to compute these weighted averages are called *soft counts* (as opposed to the *hard counts* used in a hard-EM-type algorithm such as k-means). The probabilities computed for $\mathbf{Z}$ are posterior probabilities and are what is computed in the E step. The soft counts used to compute new parameter values are what is computed in the M step.

Properties

Speaking of an expectation (E) step is a bit of a misnomer. What is calculated in the first step are the fixed, data-dependent parameters of the function Q. Once the parameters of Q are known, it is fully determined and is maximized in the second (M) step of an EM algorithm.

Although an EM iteration does increase the observed data (i.e., marginal) likelihood function, no guarantee exists that the sequence converges to a maximum likelihood estimator. For multimodal distributions, this means that an EM algorithm may converge to a local maximum of the observed data likelihood function, depending on starting values. A variety of heuristic or metaheuristic approaches exist to escape a local maximum, such as random-restart hill climbing (starting with several different random initial estimates $\theta^{(t)}$), or applying simulated annealing methods.

EM is especially useful when the likelihood is an exponential family: the E step becomes the sum of expectations of sufficient statistics, and the M step involves maximizing a linear function. In such a case, it is usually possible to derive closed-form expression updates for each step, using the Sundberg formula (published by Rolf Sundberg using unpublished results of Per Martin-Löf and Anders Martin-Löf).

The EM method was modified to compute maximum a posteriori (MAP) estimates for Bayesian inference in the original paper by Dempster, Laird, and Rubin.

Other methods exist to find maximum likelihood estimates, such as gradient descent, conjugate gradient, or variants of the Gauss–Newton algorithm. Unlike EM, such methods typically require the evaluation of first and/or second derivatives of the likelihood function.

Proof of Correctness

Expectation-maximization works to improve $Q(\theta \,|\, \theta^{(t)})$ rather than directly improving $\log p(X \,|\, \theta)$. Here is shown that improvements to the former imply improvements to the latter.

For any Z with non-zero probability $p(Z \,|\, X, \theta)$, we can write

$$\log p(X \,|\, \theta) = \log p(X, Z \,|\, \theta) - \log p(Z \,|\, X, \theta).$$

We take the expectation over possible values of the unknown data $\mathbf{Z}$ under the current parameter estimate $\theta^{(t)}$ by multiplying both sides by $p(Z \mid X, \theta^{(t)})$ and summing (or integrating) over $\mathbf{Z}$. The left-hand side is the expectation of a constant, so we get:

$$\log p(X \mid \theta) = \sum_Z p(Z \mid X, \theta^{(t)}) \log p(X, Z \mid \theta) - \sum_Z p(Z \mid X, \theta^{(t)}) \log p(Z \mid X, \theta)$$
$$= Q(\theta \mid \theta^{(t)}) + H(\theta \mid \theta^{(t)}),$$

where $H(\theta \mid \theta^{(t)})$ is defined by the negated sum it is replacing. This last equation holds for any value of θ including $\theta = \theta^{(t)}$,

$$\log p(X \mid \theta^{(t)}) = Q(\theta^{(t)} \mid \theta^{(t)}) + H(\theta^{(t)} \mid \theta^{(t)})$$

and subtracting this last equation from the previous equation gives

$$\log p(X \mid \theta) - \log p(X \mid \theta^{(t)}) = Q(\theta \mid \theta^{(t)}) - Q(\theta^{(t)} \mid \theta^{(t)}) + H(\theta \mid \theta^{(t)}) - H(\theta^{(t)} \mid \theta^{(t)})$$

However, Gibbs' inequality tells us that $H(\theta \mid \theta^{(t)}) \geq H(\theta^{(t)} \mid \theta^{(t)})$, so we can conclude that

$$\log p(X \mid \theta) - \log p(X \mid \theta^{(t)}) \geq Q(\theta \mid \theta^{(t)}) - Q(\theta^{(t)} \mid \theta^{(t)}).$$

In words, choosing θ to improve $Q(\theta \mid \theta^{(t)})$ beyond $Q(\theta^{(t)} \mid \theta^{(t)})$ can not cause $\log p(X \mid \theta)$ to decrease below $\log p(X \mid \theta^{(t)})$, and so the marginal likelihood of the data is non-decreasing.

As a Maximization-maximization Procedure

The EM algorithm can be viewed as two alternating maximization steps, that is, as an example of coordinate ascent. Consider the function:

$$F(q, \theta) := E_q[\log L(\theta; x, Z)] + H(q),$$

where q is an arbitrary probability distribution over the unobserved data z and $H(q)$ is the entropy of the distribution q. This function can be written as

$$F(q, \theta) = -D_{KL}\left(q \,\middle\|\, p_{Z|X}(\cdot \mid x; \theta)\right) + \log L(\theta; x),$$

where $p_{Z|X}(\cdot \mid x; \theta)$ is the conditional distribution of the unobserved data given the observed data x and D_{KL} is the Kullback–Leibler divergence.

Then the steps in the EM algorithm may be viewed as:

Expectation step: Choose q to maximize F:

$$q^{(t)} = \operatorname{argmax}_q F(q, \theta^{(t)})$$

Maximization step: Choose θ to maximize F:

$$\theta^{(t+1)} = \operatorname{argmax}_\theta F(q^{(t)}, \theta)$$

Applications

EM is frequently used for data clustering in machine learning and computer vision. In natural language processing, two prominent instances of the algorithm are the Baum-Welch algorithm for hidden Markov models, and the inside-outside algorithm for unsupervised induction of probabilistic context-free grammars.

In psychometrics, EM is almost indispensable for estimating item parameters and latent abilities of item response theory models.

With the ability to deal with missing data and observe unidentified variables, EM is becoming a useful tool to price and manage risk of a portfolio.

The EM algorithm (and its faster variant ordered subset expectation maximization) is also widely used in medical image reconstruction, especially in positron emission tomography and single photon emission computed tomography.

In structural engineering, the Structural Identification using Expectation Maximization (STRIDE) algorithm is an output-only method for identifying natural vibration properties of a structural system using sensor data.

Filtering and Smoothing EM Algorithms

A Kalman filter is typically used for on-line state estimation and a minimum-variance smoother may be employed for off-line or batch state estimation. However, these minimum-variance solutions require estimates of the state-space model parameters. EM algorithms can be used for solving joint state and parameter estimation problems.

Filtering and smoothing EM algorithms arise by repeating this two-step procedure:

E-step

> Operate a Kalman filter or a minimum-variance smoother designed with current parameter estimates to obtain updated state estimates.

M-step

> Use the filtered or smoothed state estimates within maximum-likelihood calculations to obtain updated parameter estimates.

Suppose that a Kalman filter or minimum-variance smoother operates on measurements of a single-input-single-output system that possess additive white noise. An updated measurement noise variance estimate can be obtained from the maximum likelihood calculation

$$\hat{\sigma}_v^2 = \frac{1}{N} \sum_{k=1}^{N} (z_k - \hat{x}_k)^2$$

where $\hat{x}_k$ are scalar output estimates calculated by a filter or a smoother from N scalar measurements z_k. The above update can also be applied to updating a Poisson measurement noise intensity. Similarly, for a first-order auto-regressive process, an updated process noise variance estimate can be calculated by

$$\hat{\sigma}_w^2 = \frac{1}{N} \sum_{k=1}^{N} (\hat{x}_{k+1} - \hat{F}\hat{x}_k)^2$$

where $\hat{x}_k$ and $\hat{x}_{k+1}$ are scalar state estimates calculated by a filter or a smoother. The updated model coefficient estimate is obtained via

$$\hat{F} = \frac{\sum_{k=1}^{N} \left(\hat{x}_{k+1} - \hat{F}''\hat{x}_k \right)}{\sum_{k=1}^{N} \hat{x}_k^2}.$$

The convergence of parameter estimates such as those above are well studied.

Variants

A number of methods have been proposed to accelerate the sometimes slow convergence of the EM algorithm, such as those using conjugate gradient and modified Newton's methods (Newton–Raphson). Also, EM can be used with constrained estimation methods.

Expectation conditional maximization (ECM) replaces each M step with a sequence of conditional maximization (CM) steps in which each parameter θ_i is maximized individually, conditionally on the other parameters remaining fixed.

This idea is further extended in *generalized expectation maximization (GEM)* algorithm, in which is sought only an increase in the objective function F for both the E step and M step as described in the As a maximization-maximization procedure section. GEM is further developed in a distributed environment and shows promising results.

It is also possible to consider the EM algorithm as a subclass of the MM (Majorize/Minimize or Minorize/Maximize, depending on context) algorithm, and therefore use any machinery developed in the more general case.

α-EM Algorithm

The Q-function used in the EM algorithm is based on the log likelihood. Therefore, it is regarded as the log-EM algorithm. The use of the log likelihood can be generalized to that of the α-log likelihood ratio. Then, the α-log likelihood ratio of the observed data can be exactly expressed as equality by using the Q-function of the α-log likelihood ratio and the α-divergence. Obtaining this Q-function is a generalized E step. Its maximization is a generalized M step. This pair is called the α-EM algorithm which contains the log-EM algorithm as its subclass. Thus, the α-EM algorithm by Yasuo Matsuyama is an exact generalization of the log-EM algorithm. No computation of gradient or Hessian matrix is needed. The α-EM shows faster convergence than the log-EM algorithm by choosing an appropriate α. The α-EM algorithm leads to a faster version of the Hidden Markov model estimation algorithm α-HMM.

Relation to Variational Bayes Methods

EM is a partially non-Bayesian, maximum likelihood method. Its final result gives a probability distribution over the latent variables (in the Bayesian style) together with a point estimate for θ (either a maximum likelihood estimate or a posterior mode). A fully Bayesian version of this may be wanted, giving a probability distribution over θ and the latent variables. The Bayesian approach to inference is simply to treat θ as another latent variable. In this paradigm, the distinction between the E and M steps disappears. If using the factorized Q approximation as described above (variational Bayes), solving can iterate over each latent variable (now including θ) and optimize them one at a time. Now, k steps per iteration are needed, where k is the number of latent variables. For graphical models this is easy to do as each variable's new Q depends only on its Markov blanket, so local message passing can be used for efficient inference.

Geometric Interpretation

In information geometry, the E step and the M step are interpreted as projections under dual affine connections, called the e-connection and the m-connection; the Kullback–Leibler divergence can also be understood in these terms.

Examples

Gaussian Mixture

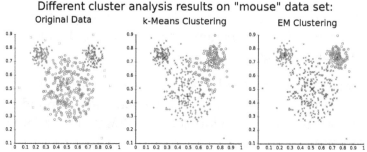

Comparison of k-means and EM on artificial Data visualized with ELKI. Using the Variances, the EM algorithm can describe the normal distributions exact, while k-Means splits the data in Voronoi-Cells. The Cluster center is visualized by the lighter, bigger Symbol.

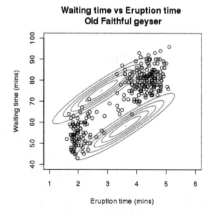

Demonstration of the EM algorithm fitting a two component Gaussian mixture model to the Old Faithful dataset. The algorithm steps through from a random initialization to convergence.

Let $\mathbf{x} = (\mathbf{x}_1, \mathbf{x}_2, \ldots, \mathbf{x}_n)$ be a sample of n independent observations from a mixture of two multivariate normal distributions of dimension d, and let $\mathbf{z} = (z_1, z_2, \ldots, z_n)$ be the latent variables that determine the component from which the observation originates.

$X_i \mid (Z_i = 1) \sim \mathcal{N}_d(\mu_1, \Sigma_1)$ and $X_i \mid (Z_i = 2) \sim \mathcal{N}_d(\mu_2, \Sigma_2)$

where

$P(Z_i = 1) = \tau_1$ and $P(Z_i = 2) = \tau_2 = 1 - \tau_1$

The aim is to estimate the unknown parameters representing the *mixing* value between the Gaussians and the means and covariances of each:

$$\theta = \left(\tau, \mu_1, \mu_2, \Sigma_1, \Sigma_2 \right)$$

where the incomplete-data likelihood function is

$$L(\theta; x) = \prod_{i=1}^{n} \sum_{j=1}^{2} \tau_j \, f(x_i; \mu_j, \Sigma_j),$$

and the complete-data likelihood function is

$$L(\theta; x, z) = p(x, z \mid \theta) = \prod_{i=1}^{n} \prod_{j=1}^{2} \left[f(x_i; \mu_j, \Sigma_j) \tau_j \right]^{\mathbb{I}(z_i = j)}$$

or

$$L(\theta; x, z) = \exp \left\{ \sum_{i=1}^{n} \sum_{j=1}^{2} \mathbb{I}(z_i = j) \left[\log \tau_j - \tfrac{1}{2} \log |\Sigma_j| - \tfrac{1}{2} (x_i - \mu_j)^{\top} \Sigma_j^{-1} (x_i - \mu_j) - \tfrac{d}{2} \log(2\pi) \right] \right\}.$$

where $\mathbb{I}$ is an indicator function and f is the probability density function of a multivariate normal.

To see the last equality, then for each *i* all indicators $\mathbb{I}(z_i = j)$ are equal to zero, except for one which is equal to one. The inner sum thus reduces to one term.

E Step

Given our current estimate of the parameters $\theta^{(t)}$, the conditional distribution of the Z_i is determined by Bayes theorem to be the proportional height of the normal density weighted by τ:

$$T_{j,i}^{(t)} := P(Z_i = j \mid X_i = x_i; \theta^{(t)}) = \frac{\tau_j^{(t)} \, f(x_i; \mu_j^{(t)}, \Sigma_j^{(t)})}{\tau_1^{(t)} \, f(x_i; \mu_1^{(t)}, \Sigma_1^{(t)}) + \tau_2^{(t)} \, f(x_i; \mu_2^{(t)}, \Sigma_2^{(t)})}.$$

These are called the "membership probabilities" which are normally considered the output of the E step (although this is not the Q function of below).

This E step corresponds with this function for Q:

$$Q(\theta \mid \theta^{(t)}) = E_{Z \mid X, \theta^{(t)}} [\log L(\theta; x, Z)]$$

$$= E_{Z \mid X, \theta^{(t)}} [\log \prod_{i=1}^{n} L(\theta; x_i, z_i)]$$

$$= E_{Z|X,\theta^{(t)}}\left[\sum_{i=1}^{n} \log L(\theta; x_i, z_i)\right]$$

$$= \sum_{i=1}^{n} E_{Z|X;\theta^{(t)}}[\log L(\theta; x_i, z_i)]$$

$$= \sum_{i=1}^{n}\sum_{j=1}^{2} P(Z_i = j \mid X_i = \mathbf{x}_i; \theta^{(t)}) \log L(\theta_j; \mathbf{x}_i, \mathbf{z}_i)$$

$$= \sum_{i=1}^{n}\sum_{j=1}^{2} T_{j,i}^{(t)}\left[\log \tau_j - \tfrac{1}{2}\log|\Sigma_j| - \tfrac{1}{2}(\mathbf{x}_i - \mu_j)^\top \Sigma_j^{-1}(\mathbf{x}_i - \mu_j) - \tfrac{d}{2}\log(2\pi)\right]$$

This full conditional expectation does not need to be calculated in one step, because τ and μ/Σ appear in separate linear terms and can thus be maximized independently.

M step

$Q(\theta|\theta^{(t)})$ being quadratic in form means that determining the maximizing values of θ is relatively straightforward. Also, τ, (μ_1,Σ_1) and (μ_2,Σ_2) may all be maximized independently since they all appear in separate linear terms.

To begin, consider τ, which has the constraint $\tau_1 + \tau_2 = 1$:

$$\tau^{(t+1)} = \arg\max_{\tau} Q(\theta \mid \theta^{(t)})$$

$$= \arg\max_{\tau}\left\{\left[\sum_{i=1}^{n} T_{1,i}^{(t)}\right]\log \tau_1 + \left[\sum_{i=1}^{n} T_{2,i}^{(t)}\right]\log \tau_2\right\}$$

This has the same form as the MLE for the binomial distribution, so

$$\tau_j^{(t+1)} = \frac{\sum_{i=1}^{n} T_{j,i}^{(t)}}{\sum_{i=1}^{n}(T_{1,i}^{(t)} + T_{2,i}^{(t)})} = \frac{1}{n}\sum_{i=1}^{n} T_{j,i}^{(t)}.$$

For the next estimates of (μ_1,Σ_1):

$$(\mu_1^{(t+1)}, \Sigma_1^{(t+1)}) = \arg\max_{\mu_1,\Sigma_1} Q(\theta \mid \theta^{(t)})$$

$$= \arg\max_{\mu_1,\Sigma_1} \sum_{i=1}^{n} T_{1,i}^{(t)}\left\{-\tfrac{1}{2}\log|\Sigma_1| - \tfrac{1}{2}(\mathbf{x}_i - \mu_1)^\top \Sigma_1^{-1}(\mathbf{x}_i - \mu_1)\right\}.$$

This has the same form as a weighted MLE for a normal distribution, so

$$\mu_2^{(t+1)} = \frac{\sum_{i=1}^{n} T_{2,i}^{(t)} x_i}{\sum_{i=1}^{n} T_{2,i}^{(t)}} \quad \text{and} \quad \Sigma_1^{(t+1)} = \frac{\sum_{i=1}^{n} T_{1,i}^{(t)}(x_i - \mu_1^{(t+1)})(x_i - \mu_1^{(t+1)})^\top}{\sum_{i=1}^{n} T_{1,i}^{(t)}}$$

and, by symmetry

$$\mu_2^{(t+1)} = \frac{\sum_{i=1}^{n} T_{2,i}^{(t)} x_i}{\sum_{i=1}^{n} T_{2,i}^{(t)}} \text{ and } \Sigma_2^{(t+1)} = \frac{\sum_{i=1}^{n} T_{2,i}^{(t)} (x_i - \mu_2^{(t+1)})(x_i - \mu_2^{(t+1)})^\top}{\sum_{i=1}^{n} T_{2,i}^{(t)}}.$$

Termination

Conclude the iterative process if $E_{Z|\theta^{(t)},x}[\log L(\theta^{(t)}; x, Z)] \leq E_{Z|\theta^{(t-1)},x}[\log L(\theta^{(t-1)}; x, Z)] + \epsilon$ for ϵ below some preset threshold.

Generalization

The algorithm illustrated above can be generalized for mixtures of more than two multivariate normal distributions.

Truncated and Censored Regression

The EM algorithm has been implemented in the case where an underlying linear regression model exists explaining the variation of some quantity, but where the values actually observed are censored or truncated versions of those represented in the model. Special cases of this model include censored or truncated observations from one normal distribution.

Alternatives

EM typically converges to a local optimum, not necessarily the global optimum, with no bound on the convergence rate in general. It is possible that it can be arbitrarily poor in high dimensions and there can be an exponential number of local optima. Hence, a need exists for alternative methods for guaranteed learning, especially in the high-dimensional setting. Alternatives to EM exist with better guarantees for consistency, which are termed *moment-based approaches* or the so-called *spectral techniques*. Moment-based approaches to learning the parameters of a probabilistic model are of increasing interest recently since they enjoy guarantees such as global convergence under certain conditions unlike EM which is often plagued by the issue of getting stuck in local optima. Algorithms with guarantees for learning can be derived for a number of important models such as mixture models, HMMs etc. For these spectral methods, no spurious local optima occur, and the true parameters can be consistently estimated under some regularity conditions.

Forward–backward Algorithm

The forward–backward algorithm is an inference algorithm for hidden Markov models which computes the posterior marginals of all hidden state variables given a sequence of observations/emissions $o_{1:t} := o_1, \dots, o_t$, i.e. it computes, for all hidden state variables $X_k \in \{X_1, \dots, X_t\}$, the distribution $P(X_k \mid o_{1:t})$. This inference task is usually called *smoothing*. The algorithm makes use of the principle of dynamic programming to compute efficiently the values that are required to obtain the posterior marginal distributions in two passes. The first pass goes forward in time while the second goes backward in time; hence the name *forward–backward algorithm*.

The term *forward–backward algorithm* is also used to refer to any algorithm belonging to the general class of algorithms that operate on sequence models in a forward–backward manner.

Overview

In the first pass, the forward–backward algorithm computes a set of forward probabilities which provide, for all $k \in \{1,\ldots,t\}$, the probability of ending up in any particular state given the first k observations in the sequence, i.e. $P(X_k \mid o_{1:k})$. In the second pass, the algorithm computes a set of backward probabilities which provide the probability of observing the remaining observations given any starting point k, i.e. $P(o_{k+1:t} \mid X_k)$. These two sets of probability distributions can then be combined to obtain the distribution over states at any specific point in time given the entire observation sequence:

$$P(X_k \mid o_{1:t}) = P(X_k \mid o_{1:k}, o_{k+1:t}) \propto P(o_{k+1:t} \mid X_k)P(o_{1:k}, X_k)$$

The last step follows from an application of the Bayes' rule and the conditional independence of $o_{k+1:t}$ and $o_{1:k}$ given X_k.

As outlined above, the algorithm involves three steps:

1. computing forward probabilities

2. computing backward probabilities

3. computing smoothed values.

The forward and backward steps may also be called "forward message pass" and "backward message pass" - these terms are due to the *message-passing* used in general belief propagation approaches. At each single observation in the sequence, probabilities to be used for calculations at the next observation are computed. The smoothing step can be calculated simultaneously during the backward pass. This step allows the algorithm to take into account any past observations of output for computing more accurate results.

The forward–backward algorithm can be used to find the most likely state for any point in time. It cannot, however, be used to find the most likely sequence of states.

Forward Probabilities

The following description will use matrices of probability values rather than probability distributions, although in general the forward-backward algorithm can be applied to continuous as well as discrete probability models.

We transform the probability distributions related to a given hidden Markov model into matrix notation as follows. The transition probabilities $P(X_t \mid X_{t-1})$ of a given random variable X_t representing all possible states in the hidden Markov model will be represented by the matrix $\mathbf{T}$

where the column index i will represent the target state and the row index j represents the start state. A transition from row-vector state π_t to the incremental row-vector state π_{t+1} is written as $\pi_{t+1} = \pi_t T$. The example below represents a system where the probability of staying in the same state after each step is 70% and the probability of transitioning to the other state is 30%. The transition matrix is then:

$$\mathbf{T} = \begin{pmatrix} 0.7 & 0.3 \\ 0.3 & 0.7 \end{pmatrix}$$

In a typical Markov model we would multiply a state vector by this matrix to obtain the probabilities for the subsequent state. In a hidden Markov model the state is unknown, and we instead observe events associated with the possible states. An event matrix of the form:

$$\mathbf{B} = \begin{pmatrix} 0.9 & 0.1 \\ 0.2 & 0.8 \end{pmatrix}$$

provides the probabilities for observing events given a particular state. In the above example, event 1 will be observed 90% of the time if we are in state 1 while event 2 has a 10% probability of occurring in this state. In contrast, event 1 will only be observed 20% of the time if we are in state 2 and event 2 has an 80% chance of occurring. Given an arbitrary row-vector describing the state of the system (π), the probability of observing event j is then:

$$P(O = j) = \sum_i \pi_i b_{j,i}$$

This can be represented in matrix form by multiplying the state row-vector (π) by an observation matrix ($\mathbf{O_j} = \mathrm{diag}(b_{*,o_j})$) containing only diagonal entries. Each entry is the probability of the observed event given each state. Continuing the above example, an observation of tevent 1 would be:

$$\mathbf{O_1} = \begin{pmatrix} 0.9 & 0.0 \\ 0.0 & 0.2 \end{pmatrix}$$

This allows us to calculate the new unnormalized probabilities state vector π' through Bayes rule, weighting by the likelihood that each element of π generated event 1 as:

$$\pi' = \pi \mathbf{O_1}$$

We can now make this general procedure specific to our series of observations. Assuming an initial state vector π_0, (which can be optimized as a parameter through repetitions of the forward-back procedure), we begin with:

$$f_{0:0} = \pi_0 \mathbf{T} \mathbf{O}_{o(0)}$$

This process can be carried forward with additional observations using:

$$\mathbf{f}_{0:t} = \mathbf{f}_{0:t-1}\mathbf{TO}_{o(t)}$$

This value is the forward unnormalized probability vector. The i'th entry of this vector provides:

$$f_{0:t}(i) = P(o_1, o_2, \ldots, o_t, X_t = x_i \mid \pi)$$

Typically, we will normalize the probability vector at each step so that its entries sum to 1. A scaling factor is thus introduced at each step such that:

$$\hat{\mathbf{f}}_{0:t} = c_t^{-1}\,\mathbf{f}_{0:t-1}\mathbf{TO}_{o(t)}$$

where $\hat{\mathbf{f}}_{0:t-1}$ represents the scaled vector from the previous step and c_t represents the scaling factor that causes the resulting vector's entries to sum to 1. The product of the scaling factors is the total probability for observing the given events irrespective of the final states:

$$P(o_1, o_2, \ldots, o_t \mid \pi) = \prod_{s=1}^{t} c_s$$

This allows us to interpret the scaled probability vector as:

$$\hat{f}_{0:t}(i) = \frac{f_{0:t}(i)}{\displaystyle\prod_{s=1}^{t} c_s} = \frac{P(o_1, o_2, \ldots, o_t, X_t = x_i \mid \pi)}{P(o_1, o_2, \ldots, o_t \mid \pi)} = P(X_t = x_i \mid o_1, o_2, \ldots, o_t, \pi)$$

We thus find that the product of the scaling factors provides us with the total probability for observing the given sequence up to time t and that the scaled probability vector provides us with the probability of being in each state at this time.

Backward Probabilities

A similar procedure can be constructed to find backward probabilities. These intend to provide the probabilities:

$$\mathbf{b}_{t:T}(i) = P(o_{t+1}, o_{t+2}, \ldots, o_T \mid X_t = x_i)$$

That is, we now want to assume that we start in a particular state ($X_t = x_i$), and we are now interested in the probability of observing all future events from this state. Since the initial state is assumed as given (i.e. the prior probability of this state = 100%), we begin with:

$$\mathbf{b}_{T:T} = [1\,1\,1\ldots]^{\mathrm{T}}$$

Notice that we are now using a column vector while the forward probabilities used row vectors. We can then work backwards using:

$$\mathbf{b}_{t-1:T} = \mathbf{TO}_t\mathbf{b}_{t:T}$$

While we could normalize this vector as well so that its entries sum to one, this is not usually done. Noting that each entry contains the probability of the future event sequence given a particular initial state, normalizing this vector would be equivalent to applying Bayes' theorem to find the likelihood of each initial state given the future events (assuming uniform priors for the final state vector). However, it is more common to scale this vector using the same c_t constants used in the forward probability calculations. $\mathbf{b}_{T:T}$ is not scaled, but subsequent operations use:

$$\hat{\mathbf{b}}_{t-1:T} = c_t^{-1}\mathbf{TO}_t\hat{\mathbf{b}}_{t:T}$$

where $\hat{\mathbf{b}}_{t:T}$ represents the previous, scaled vector. This result is that the scaled probability vector is related to the backward probabilities by:

$$\hat{\mathbf{b}}_{t:T}(i) = \frac{\mathbf{b}_{t:T}(i)}{\displaystyle\prod_{s=t+1}^{T} c_s}$$

This is useful because it allows us to find the total probability of being in each state at a given time, t, by multiplying these values:

$$\gamma_t(i) = P(X_t = x_i \mid o_1, o_2, \ldots, o_T, \pi) = \frac{P(o_1, o_2, \ldots, o_T, X_t = x_i \mid \pi)}{P(o_1, o_2, \ldots, o_T \mid \pi)} = \frac{f_{0:t}(i) \cdot b_{t:T}(i)}{\displaystyle\prod_{s=1}^{T} c_s} = \hat{f}_{0:t}(i) \cdot \hat{b}_{t:T}(i)$$

To understand this, we note that $\mathbf{f}_{0:t}(i) \cdot \mathbf{b}_{t:T}(i)$ provides the probability for observing the given events in a way that passes through state x_i at time t. This probability includes the forward probabilities covering all events up to time t as well as the backward probabilities which include all future events. This is the numerator we are looking for in our equation, and we divide by the total probability of the observation sequence to normalize this value and extract only the probability that $X_t = x_i$. These values are sometimes called the "smoothed values" as they combine the forward and backward probabilities to compute a final probability.

The values $\gamma_t(i)$ thus provide the probability of being in each state at time t. As such, they are useful for determining the most probable state at any time. It should be noted, however, that the term "most probable state" is somewhat ambiguous. While the most probable state is the most likely to be correct at a given point, the sequence of individually probable states is not likely to be the most probable sequence. This is because the probabilities for each point are calculated independently of each other. They do not take into account the transition probabilities between states, and it is thus possible to get states at two moments (t and t+1) that are both most probable at those time points but which have very little probability of occurring together, i.e. $P(X_t = x_i, X_{t+1} = x_j) \neq P(X_t = x_i)P(X_{t+1} = x_j)$. The most probable sequence of states that produced an observation sequence can be found using the Viterbi algorithm.

Example

This example takes as its basis the umbrella world in Russell & Norvig 2010 Chapter 15 pp. 566 in which we would like to infer the weather given observation of a man either carrying or not carrying an umbrella. We assume two possible states for the weather: state 1 = rain, state 2 = no rain. We assume that the weather has a 70% chance of staying the same each day and a 30% chance of changing. The transition probabilities are then:

$$\mathbf{T} = \begin{pmatrix} 0.7 & 0.3 \\ 0.3 & 0.7 \end{pmatrix}$$

We also assume each state generates 2 events: event 1 = umbrella, event 2 = no umbrella. The conditional probabilities for these occurring in each state are given by the probability matrix:

$$\mathbf{B} = \begin{pmatrix} 0.9 & 0.1 \\ 0.2 & 0.8 \end{pmatrix}$$

We then observe the following sequence of events: {umbrella, umbrella, no umbrella, umbrella, umbrella} which we will represent in our calculations as:

$$\mathbf{O}_1 = \begin{pmatrix} 0.9 & 0.0 \\ 0.0 & 0.2 \end{pmatrix} \mathbf{O}_2 = \begin{pmatrix} 0.9 & 0.0 \\ 0.0 & 0.2 \end{pmatrix} \mathbf{O}_3 = \begin{pmatrix} 0.1 & 0.0 \\ 0.0 & 0.8 \end{pmatrix} \mathbf{O}_4 = \begin{pmatrix} 0.9 & 0.0 \\ 0.0 & 0.2 \end{pmatrix} \mathbf{O}_5 = \begin{pmatrix} 0.9 & 0.0 \\ 0.0 & 0.2 \end{pmatrix}$$

Note that $\mathbf{O}_3$ differs from the others because of the "no umbrella" observation.

In computing the forward probabilities we begin with:

$$\mathbf{f}_{0:0} = \begin{pmatrix} 0.5 & 0.5 \end{pmatrix}$$

which is our prior state vector indicating that we don't know which state the weather is in before our observations. While a state vector should be given as a row vector, we will use the transpose of the matrix so that the calculations below are easier to read. Our calculations are then written in the form:

$$(\hat{\mathbf{f}}_{0:t})^{\mathrm{T}} = c^{-1}\mathbf{O}_t(\mathbf{T})^{\mathrm{T}}(\hat{\mathbf{f}}_{0:t-1})^{\mathrm{T}}$$

instead of:

$$\hat{\mathbf{f}}_{0:t} = c^{-1}\hat{\mathbf{f}}_{0:t-1}\mathbf{T}\mathbf{O}_t$$

Notice that the transformation matrix is also transposed, but in our example the transpose is equal to the original matrix. Performing these calculations and normalizing the results provides:

$$(\hat{\mathbf{f}}_{0:1})^{\mathrm{T}} = c_1^{-1}\begin{pmatrix} 0.9 & 0.0 \\ 0.0 & 0.2 \end{pmatrix}\begin{pmatrix} 0.7 & 0.3 \\ 0.3 & 0.7 \end{pmatrix}\begin{pmatrix} 0.5000 \\ 0.5000 \end{pmatrix} = c_1^{-1}\begin{pmatrix} 0.4500 \\ 0.1000 \end{pmatrix} = \begin{pmatrix} 0.8182 \\ 0.1818 \end{pmatrix}$$

$$(\hat{\mathbf{f}}_{0:2})^{\mathrm{T}} = c_2^{-1}\begin{pmatrix} 0.9 & 0.0 \\ 0.0 & 0.2 \end{pmatrix}\begin{pmatrix} 0.7 & 0.3 \\ 0.3 & 0.7 \end{pmatrix}\begin{pmatrix} 0.8182 \\ 0.1818 \end{pmatrix} = c_2^{-1}\begin{pmatrix} 0.5645 \\ 0.0745 \end{pmatrix} = \begin{pmatrix} 0.8834 \\ 0.1166 \end{pmatrix}$$

$$(\hat{\mathbf{f}}_{0:3})^{\mathrm{T}} = c_3^{-1} \begin{pmatrix} 0.1 & 0.0 \\ 0.0 & 0.8 \end{pmatrix} \begin{pmatrix} 0.7 & 0.3 \\ 0.3 & 0.7 \end{pmatrix} \begin{pmatrix} 0.8834 \\ 0.1166 \end{pmatrix} = c_3^{-1} \begin{pmatrix} 0.0653 \\ 0.2772 \end{pmatrix} = \begin{pmatrix} 0.1907 \\ 0.8093 \end{pmatrix}$$

$$(\hat{\mathbf{f}}_{0:4})^{\mathrm{T}} = c_4^{-1} \begin{pmatrix} 0.9 & 0.0 \\ 0.0 & 0.2 \end{pmatrix} \begin{pmatrix} 0.7 & 0.3 \\ 0.3 & 0.7 \end{pmatrix} \begin{pmatrix} 0.1907 \\ 0.8093 \end{pmatrix} = c_4^{-1} \begin{pmatrix} 0.3386 \\ 0.1247 \end{pmatrix} = \begin{pmatrix} 0.7308 \\ 0.2692 \end{pmatrix}$$

$$(\hat{\mathbf{f}}_{0:5})^{\mathrm{T}} = c_5^{-1} \begin{pmatrix} 0.9 & 0.0 \\ 0.0 & 0.2 \end{pmatrix} \begin{pmatrix} 0.7 & 0.3 \\ 0.3 & 0.7 \end{pmatrix} \begin{pmatrix} 0.7308 \\ 0.2692 \end{pmatrix} = c_5^{-1} \begin{pmatrix} 0.5331 \\ 0.0815 \end{pmatrix} = \begin{pmatrix} 0.8673 \\ 0.1327 \end{pmatrix}$$

For the backward probabilities we start with:

$$\mathbf{b}_{5:5} = \begin{pmatrix} 1.0 \\ 1.0 \end{pmatrix}$$

We are then able to compute (using the observations in reverse order and normalizing with different constants):

$$\hat{\mathbf{b}}_{4:5} = \alpha \begin{pmatrix} 0.7 & 0.3 \\ 0.3 & 0.7 \end{pmatrix} \begin{pmatrix} 0.9 & 0.0 \\ 0.0 & 0.2 \end{pmatrix} \begin{pmatrix} 1.0000 \\ 1.0000 \end{pmatrix} = \alpha \begin{pmatrix} 0.6900 \\ 0.4100 \end{pmatrix} = \begin{pmatrix} 0.6273 \\ 0.3727 \end{pmatrix}$$

$$\hat{\mathbf{b}}_{3:5} = \alpha \begin{pmatrix} 0.7 & 0.3 \\ 0.3 & 0.7 \end{pmatrix} \begin{pmatrix} 0.9 & 0.0 \\ 0.0 & 0.2 \end{pmatrix} \begin{pmatrix} 0.6273 \\ 0.3727 \end{pmatrix} = \alpha \begin{pmatrix} 0.4175 \\ 0.2215 \end{pmatrix} = \begin{pmatrix} 0.6533 \\ 0.3467 \end{pmatrix}$$

$$\hat{\mathbf{b}}_{2:5} = \alpha \begin{pmatrix} 0.7 & 0.3 \\ 0.3 & 0.7 \end{pmatrix} \begin{pmatrix} 0.1 & 0.0 \\ 0.0 & 0.8 \end{pmatrix} \begin{pmatrix} 0.6533 \\ 0.3467 \end{pmatrix} = \alpha \begin{pmatrix} 0.1289 \\ 0.2138 \end{pmatrix} = \begin{pmatrix} 0.3763 \\ 0.6237 \end{pmatrix}$$

$$\hat{\mathbf{b}}_{1:5} = \alpha \begin{pmatrix} 0.7 & 0.3 \\ 0.3 & 0.7 \end{pmatrix} \begin{pmatrix} 0.9 & 0.0 \\ 0.0 & 0.2 \end{pmatrix} \begin{pmatrix} 0.3763 \\ 0.6237 \end{pmatrix} = \alpha \begin{pmatrix} 0.2745 \\ 0.1889 \end{pmatrix} = \begin{pmatrix} 0.5923 \\ 0.4077 \end{pmatrix}$$

$$\hat{\mathbf{b}}_{0:5} = \alpha \begin{pmatrix} 0.7 & 0.3 \\ 0.3 & 0.7 \end{pmatrix} \begin{pmatrix} 0.9 & 0.0 \\ 0.0 & 0.2 \end{pmatrix} \begin{pmatrix} 0.5923 \\ 0.4077 \end{pmatrix} = \alpha \begin{pmatrix} 0.3976 \\ 0.2170 \end{pmatrix} = \begin{pmatrix} 0.6469 \\ 0.3531 \end{pmatrix}$$

Finally, we will compute the smoothed probability values. These result also must be scaled so that its entries sum to 1 because we did not scale the backward probabilities with the c_t's found earlier. The backward probability vectors above thus actually represent the likelihood of each state at time t given the future observations. Because these vectors are proportional to the actual backward probabilities, the result has to be scaled an additional time.

$$(\gamma_0)^{\mathrm{T}} = \alpha \begin{pmatrix} 0.5000 \\ 0.5000 \end{pmatrix} \circ \begin{pmatrix} 0.6469 \\ 0.3531 \end{pmatrix} = \alpha \begin{pmatrix} 0.3235 \\ 0.1765 \end{pmatrix} = \begin{pmatrix} 0.6469 \\ 0.3531 \end{pmatrix}$$

$$(\gamma_1)^{\mathrm{T}} = \alpha \begin{pmatrix} 0.8182 \\ 0.1818 \end{pmatrix} \circ \begin{pmatrix} 0.5923 \\ 0.4077 \end{pmatrix} = \alpha \begin{pmatrix} 0.4846 \\ 0.0741 \end{pmatrix} = \begin{pmatrix} 0.8673 \\ 0.1327 \end{pmatrix}$$

$$(\gamma_2)^T = \alpha \begin{pmatrix} 0.8834 \\ 0.1166 \end{pmatrix} \circ \begin{pmatrix} 0.3763 \\ 0.6237 \end{pmatrix} = \alpha \begin{pmatrix} 0.3324 \\ 0.0728 \end{pmatrix} = \begin{pmatrix} 0.8204 \\ 0.1796 \end{pmatrix}$$

$$(\gamma_3)^T = \alpha \begin{pmatrix} 0.1907 \\ 0.8093 \end{pmatrix} \circ \begin{pmatrix} 0.6533 \\ 0.3467 \end{pmatrix} = \alpha \begin{pmatrix} 0.1246 \\ 0.2806 \end{pmatrix} = \begin{pmatrix} 0.3075 \\ 0.6925 \end{pmatrix}$$

$$(\gamma_4)^T = \alpha \begin{pmatrix} 0.7308 \\ 0.2692 \end{pmatrix} \circ \begin{pmatrix} 0.6273 \\ 0.3727 \end{pmatrix} = \alpha \begin{pmatrix} 0.4584 \\ 0.1003 \end{pmatrix} = \begin{pmatrix} 0.8204 \\ 0.1796 \end{pmatrix}$$

$$(\gamma_5)^T = \alpha \begin{pmatrix} 0.8673 \\ 0.1327 \end{pmatrix} \circ \begin{pmatrix} 1.0000 \\ 1.0000 \end{pmatrix} = \alpha \begin{pmatrix} 0.8673 \\ 0.1327 \end{pmatrix} = \begin{pmatrix} 0.8673 \\ 0.1327 \end{pmatrix}$$

Notice that the value of γ_0 is equal to $\hat{b}_{0:5}$ and that γ_5 is equal to $\hat{f}_{0:5}$. This follows naturally because both $\hat{f}_{0:5}$ and $\hat{b}_{0:5}$ begin with uniform priors over the initial and final state vectors (respectively) and take into account all of the observations. However, γ_0 will only be equal to $\hat{b}_{0:5}$ when our initial state vector represents a uniform prior (i.e. all entries are equal). When this is not the case $\hat{b}_{0:5}$ needs to be combined with the initial state vector to find the most likely initial state. We thus find that the forward probabilities by themselves are sufficient to calculate the most likely final state. Similarly, the backward probabilities can be combined with the initial state vector to provide the most probable initial state given the observations. The forward and backward probabilities need only be combined to infer the most probable states between the initial and final points.

The calculations above reveal that the most probable weather state on every day except for the third one was "rain." They tell us more than this, however, as they now provide a way to quantify the probabilities of each state at different times. Perhaps most importantly, our value at γ_5 quantifies our knowledge of the state vector at the end of the observation sequence. We can then use this to predict the probability of the various weather states tomorrow as well as the probability of observing an umbrella.

Performance

The brute-force procedure for the solution of this problem is the generation of all possible N^T state sequences and calculating the joint probability of each state sequence with the observed series of events. This approach has time complexity $O(T \cdot N^T)$, where T is the length of sequences and N is the number of symbols in the state alphabet. This is intractable for realistic problems, as the number of possible hidden node sequences typically is extremely high. However, the forward–backward algorithm has time complexity $O(N^2 T)$.

An enhancement to the general forward-backward algorithm, called the Island algorithm, trades smaller memory usage for longer running time, taking $O(N^2 T \log T)$ time and $O(N \log T)$ memory. On a computer with an unlimited number of processors, this can be reduced to $O(N^2 T)$ total time, while still taking only $O(N \log T)$ memory.

In addition, algorithms have been developed to compute $f_{0:t+1}$ efficiently through online smoothing such as the fixed-lag smoothing (FLS) algorithm Russell & Norvig 2010.

Pseudocode

```
Backward(guessState, sequenceIndex):

    if sequenceIndex is past the end of the sequence, return 1

    if (guessState, sequenceIndex) has been seen before, return saved
result

    result = 0

    for each neighboring state n:

        result = result + (transition probability from guessState to
                                n given observation element at sequenceIndex)
                    * Backward(n, sequenceIndex+1)

    save result for (guessState, sequenceIndex)

    return result
```

Python example

Given HMM (just like in Viterbi algorithm) represented in the Python programming language:

```python
states = ('Healthy', 'Fever')
end_state = 'E'

observations = ('normal', 'cold', 'dizzy')

start_probability = {'Healthy': 0.6, 'Fever': 0.4}

transition_probability = {
    'Healthy' : {'Healthy': 0.69, 'Fever': 0.3, 'E': 0.01},
    'Fever'   : {'Healthy': 0.4, 'Fever': 0.59, 'E': 0.01},
    }

emission_probability = {
    'Healthy' : {'normal': 0.5, 'cold': 0.4, 'dizzy': 0.1},
    'Fever'   : {'normal': 0.1, 'cold': 0.3, 'dizzy': 0.6},
    }
```

We can write implementation like this:

```
def fwd_bkw(observations, states, start_prob, trans_prob, emm_prob, end_
st):

    # forward part of the algorithm

    fwd = []

    f_prev = {}

    for i, observation_i in enumerate(observations):

        f_curr = {}

        for st in states:

            if i == 0:

                # base case for the forward part

                prev_f_sum = start_prob[st]

            else:

                prev_f_sum = sum(f_prev[k]*trans_prob[k][st] for k in
states)


            f_curr[st] = emm_prob[st][observation_i] * prev_f_sum


        fwd.append(f_curr)

        f_prev = f_curr


    p_fwd = sum(f_curr[k] * trans_prob[k][end_st] for k in states)


    # backward part of the algorithm

    bkw = []

    b_prev = {}

    for i, observation_i_plus in enumerate(reversed(observations[1:]+(-
None,))):

        b_curr = {}
```

```
        for st in states:

            if i == 0:

                # base case for backward part

                b_curr[st] = trans_prob[st][end_st]

            else:

                b_curr[st] = sum(trans_prob[st][l] * emm_prob[l][obser-
vation_i_plus] * b_prev[l] for l in states)


        bkw.insert(0,b_curr)

        b_prev = b_curr


    p_bkw = sum(start_prob[l] * emm_prob[l][observations] * b_curr[l] for
l in states)


    # merging the two parts

    posterior = []

    for i in range(len(observations)):

        posterior.append({st: fwd[i][st] * bkw[i][st] / p_fwd for st in
states})


    assert p_fwd == p_bkw

    return fwd, bkw, posterior
```

The function fwd_bkw takes the following arguments: x is the sequence of observations, e.g. ['normal', 'cold', 'dizzy']; states is the set of hidden states; a_0 is the start probability; a are the transition probabilities; and e are the emission probabilities.

For simplicity of code, we assume that the observation sequence x is non-empty and that a[i][j] and e[i][j] is defined for all states i,j.

In the running example, the forward-backward algorithm is used as follows:

```
def example():

    return fwd_bkw(observations,
```

```
            states,

            start_probability,

            transition_probability,

            emission_probability,

            end_state)
>>> for line in example():
...     print(*line)
...
```

{'Healthy': 0.3, 'Fever': 0.04000000000000001} {'Healthy': 0.0892, 'Fever': 0.03408} {'Healthy': 0.007518, 'Fever': 0.028120319999999997}

{'Healthy': 0.0010418399999999998, 'Fever': 0.00109578} {'Healthy': 0.00249, 'Fever': 0.00394} {'Healthy': 0.01, 'Fever': 0.01}

{'Healthy': 0.8770110375573259, 'Fever': 0.1229889624426741} {'Healthy': 0.623228030950954, 'Fever': 0.3767719690490461} {'Healthy': 0.2109527048413057, 'Fever': 0.7890472951586943}

Structured kNN

The Structured k-Nearest Neighbours is a machine learning algorithm that generalizes the k-Nearest Neighbors (kNN) classifier. Whereas the kNN classifier supports binary classification, multiclass classification and regression, the Structured kNN (SkNN) allows training of a classifier for general structured output labels.

As an example, a sample instance might be a natural language sentence, and the output label is an annotated parse tree. Training a classifier consists of showing pairs of correct sample and output label pairs. After training, the structured kNN model allows one to predict for new sample instances the corresponding output label; that is, given a natural language sentence, the classifier can produce the most likely parse tree.

Training

As a training set SkNN accepts sequences of elements with defined class labels. Type of elements does not matter, the only condition is the existence of metric function that defines a distance between each pair of elements of a set.

SkNN is based on idea of creating a graph, each node of which represents class label. There is an edge between a pair of nodes iff there is a sequence of two elements in training set with corresponding classes. Thereby the first step of SkNN training is the construction of described graph from training sequences. There are two special nodes in the graph corresponding to an end and a beginning of sentences. If sequence starts with class `C`, the edge between node `START` and node `C` should be created.

Like a regular kNN, the second part of the training of SkNN consists only of storing the elements of trained sequence in special way. Each element of training sequences is stored in node related to the class of previous element in sequence. Every first element is stored in node `START`.

Inference

Labelling of input sequences in SkNN consists in finding sequence of transitions in graph, starting from node `START`, which minimises overall cost of path. Each transition corresponds to a single element of input sequence and visa versa. As a result, label of element is determined as target node label of the transition. Cost of the path is defined as sum of all its transitions, and the cost of transition from node `A` to node `B` is a distance from current input sequence element to the nearest element of class `B`, stored in node `A`. Searching of optimal path may be performed using modified Viterbi algorithm. Unlike the original one, the modified algorithm instead of maximizing the product of probabilities minimizes the sum of the distances.

Wake-sleep Algorithm

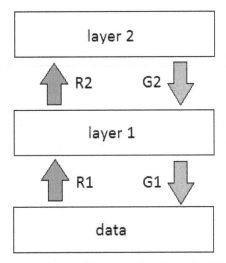

Layers of the neural network. R, G are weights used by the wake-sleep algorithm to modify data inside the layers.

The wake-sleep algorithm is an unsupervised learning algorithm for a stochastic multilayer neural network. The algorithm adjusts the parameters so as to produce a good density estimator. There are two learning phases, the "wake" phase and the "sleep" phase, which are performed alternately. It was first designed as a model for brain functioning using variational Bayesian learning. After that, the algorithm was adapted to machine learning. It can be viewed as a way to train a Helmholtz Machine

Description

The wake-sleep algorithms is visualized as a stack of layers containing representations of data.

Layers above represent data from the layer below it. Actual data is placed below the bottom layer, causing layers on top of it to become gradually more abstract. Between each pair of layers there is a recognition weight and generative weight, which are trained to improve reliability during the algorithm runtime.

The wake-sleep algorithm is convergent and can be stochastic if alternated appropriately.

Training

Training consists of two phases – the "wake" phase and the "sleep" phase.

The "Wake" Phase

Neurons are fired by recognition connections (from what would be input to what would be output). Generative connections (leading from outputs to inputs) are then modified to increase probability that they would recreate the correct activity in the layer below – closer to actual data from sensory input.

The "Sleep" Phase

The process is reversed in the "sleep" phase – neurons are fired by generative connections while recognition connections are being modified to increase probability that they would recreate the correct activity in the layer above – further to actual data from sensory input.

Potential Risks

Variational Bayesian learning is based on probabilities. There is a chance that an approximation is performed with mistakes, damaging further data representations. Another downside pertains to complicated or corrupted data samples, making it difficult to infer a representational pattern.

The wake-sleep algorithm has been suggested not to be powerful enough for the layers of the inference network in order to recover a good estimator of the posterior distribution of latent variables.

Manifold Alignment

Manifold alignment is a class of machine learning algorithms that produce projections between sets of data, given that the original data sets lie on a common manifold. The concept was first introduced as such by Ham, Lee, and Saul in 2003, adding a manifold constraint to the general problem of correlating sets of high-dimensional vectors.

Overview

Manifold alignment assumes that disparate data sets produced by similar generating processes will share a similar underlying manifold representation. By learning projections from each original space to the shared manifold, correspondences are recovered and knowledge from one domain can be transferred to another. Most manifold alignment techniques consider only two data sets, but the concept extends to arbitrarily many initial data sets.

Consider the case of aligning two data sets, X and Y, with $X_i \in \mathbb{R}^m$ and $Y_i \in \mathbb{R}^n$.

Manifold alignment algorithms attempt to project both X and Y into a new d-dimensional space such that the projections both minimize distance between corresponding points and preserve the local manifold structure of the original data. The projection functions are denoted:

$$\phi_X : \mathbb{R}^m \to \mathbb{R}^d$$

$$\phi_Y : \mathbb{R}^n \to \mathbb{R}^d$$

Let W represent the binary correspondence matrix between points in X and Y:

$$W_{i,j} = \begin{cases} 1 & \text{if } X_i \leftrightarrow Y_j \\ 0 & \text{otherwise} \end{cases}$$

Let S_X and S_Y represent pointwise similarities within data sets. This is usually encoded as the heat kernel of the adjacency matrix of a k-nearest neighbor graph.

Finally, introduce a coefficient $0 \le \mu \le 1$, which can be tuned to adjust the weight of the 'preserve manifold structure' goal, versus the 'minimize corresponding point distances' goal.

With these definitions in place, the loss function for manifold alignment can be written:

$$\arg\min_{\phi_X, \phi_Y} \mu \sum_{i,j} \left\| \phi_X(X_i) - \phi_X(X_j) \right\|^2 S_{X,i,j} + \mu \sum_{i,j} \left\| \phi_Y(Y_i) - \phi_Y(Y_j) \right\|^2 S_{Y,i,j}$$

$$+ (1-\mu) \sum_{i,j} \| \phi_X(X_i) - \phi_Y(Y_j) \|^2 W_{i,j}$$

Solving this optimization problem is equivalent to solving a generalized eigenvalue problem using the graph laplacian of the joint matrix, G:

$$G = \begin{bmatrix} \mu S_X & (1-\mu)W \\ (1-\mu)W^T & \mu S_Y \end{bmatrix}$$

Inter-data Correspondences

The algorithm described above requires full pairwise correspondence information between input data sets; a supervised learning paradigm. However, this information is usually difficult or impossible to obtain in real world applications. Recent work has extended the core manifold alignment algorithm to semi-supervised, unsupervised, and multiple-instance settings.

One-step vs. Two-step Alignment

The algorithm described above performs a "one-step" alignment, finding embeddings for both data sets at the same time. A similar effect can also be achieved with "two-step" alignments, following a slightly modified procedure:

1. Project each input data set to a lower-dimensional space independently, using any of a variety of dimension reduction algorithms.

2. Perform linear manifold alignment on the embedded data, holding the first data set fixed, mapping each additional data set onto the first's manifold. This approach has the benefit of decomposing the required computation, which lowers memory overhead and allows parallel implementations.

Instance-level vs. Feature-level Projections

Manifold alignment can be used to find linear (feature-level) projections, or nonlinear (instance-level) embeddings. While the instance-level version generally produces more accurate alignments, it sacrifices a great degree of flexibility as the learned embedding is often difficult to parameterize. Feature-level projections allow any new instances to be easily embedded in the manifold space, and projections may be combined to form direct mappings between the original data representations. These properties are especially important for knowledge-transfer applications.

Applications

Manifold alignment is suited to problems with several corpora that lie on a shared manifold, even when each corpus is of a different dimensionality. Many real-world problems fit this description, but traditional techniques are not able to take advantage of all corpora at the same time. Manifold alignment also facilitates transfer learning, in which knowledge of one domain is used to jump-start learning in correlated domains.

Applications of manifold alignment include:

- Cross-language information retrieval / automatic translation

  o By representing documents as vector of word counts, manifold alignment can recover the mapping between documents of different languages.

  o Cross-language document correspondence is relatively easy to obtain, especially from multi-lingual organizations like the European Union.

- Transfer learning of policy and state representations for reinforcement learning

- Alignment of protein NMR structures

Randomized Weighted Majority Algorithm

The randomized weighted majority algorithm is an algorithm in machine learning theory. It improves the mistake bound of the weighted majority algorithm.

Imagine that every morning before the stock market opens, we get a prediction from each of our "experts" about whether the stock market will go up or down. Our goal is to somehow combine this set of predictions into a single prediction that we then use to make a buy or sell decision for the day. The RWMA gives us a way to do this combination such that our prediction record will be nearly as good as that of the single best expert in hindsight.

Motivation

In machine learning, the weighted majority algorithm (WMA) is a meta-learning algorithm which "predicts from expert advice". It is not a randomized algorithm:

```
initialize all experts to weight 1.

for each round:

    poll all the experts and predict based on a weighted majority vote
of their predictions.

    cut in half the weights of all experts that make a mistake.
```

Suppose there are n experts and the best expert makes m mistakes. The weighted majority algorithm (WMA) makes at most $2.4(\log_2 n + m)$ mistakes, which is not a very good bound. We can do better by introducing randomization.

Randomized Weighted Majority Algorithm (RWMA)

The nonrandomized weighted majority algorithm (WMA) only guarantees an upper bound of $2.4(\log_2 n + m)$, which is problematic for highly error-prone experts (e.g. the best expert still makes a mistake 20% of the time.) Suppose we do $N = 100$ rounds using $n = 10$ experts. If the best expert makes $m = 20$ mistakes, we can only guarantee an upper bound of $2.4(\log_2 10 + 20) \approx 56$ on our number of mistakes.

As this is a known limitation of WMA, attempts to improve this shortcoming have been explored in order to improve the dependence on m. Instead of predicting based on majority vote, the weights are used as probabilities: hence the name randomized weighted majority. If w_i is the weight of expert i, let $W = \sum_i w_i$. We will follow expert i with probability $\dfrac{w_i}{W}$. The goal is to bound the worst-case expected number of mistakes, assuming that the adversary (the world) has to select one of the answers as correct before we make our coin toss. Why is this better in the worst case? Idea: the worst case for the deterministic algorithm (weighted majority algorithm) was when the weights split 50/50. But, now it is not so bad since we also have a 50/50 chance of getting it right. Also, to trade-off between dependence on m and $\log_2 n$, we will generalize to multiply by $\beta < 1$, instead of necessarily by $\dfrac{1}{2}$.

Analysis

At the t-th round, define F_t to be the fraction of weight on the wrong answers. so, F_t is the probability we make a mistake on the t-th round. Let M denote the total number of mistakes we made so far. Furthermore, we define $E[M] = \sum_t F_t$, using the fact that expectation is additive. On the t

-th round, W becomes $W(1-(1-\beta)F_t)$. Reason: on F_t fraction, we are multiplying by β. So, $W_{final} = n*(1-(1-\beta)F_1)*(1-(1-\beta)F_2)...$

Let's say that m is the number of mistakes of the best expert so far. We can use the inequality $W \geq \beta^m$. Now we solve. First, take the natural log of both sides. We get: $m\ln\beta \leq \ln(n) + \sum_t \ln(1-(1-\beta)F_t)$, Simplify:

$$\ln(1-x) = -x - \frac{x^2}{2} - \frac{x^3}{3} - ..., \text{ So,}$$

$$\ln(1-(1-\beta)F_t) < -(1-\beta)F_t.$$

$$m\ln\beta \leq \ln(n) - (1-\beta)*\sum_t F_t$$

Now, use $E[M] = \sum_t F_t$, and the result is:

$$E[M] \leq \frac{m\ln(1/\beta) + \ln(n)}{1-\beta}$$

Let's see if we made any progress:

If $\beta = \frac{1}{2}$, we get, $1.39m + 2\ln(n).,$

if $\beta = \frac{3}{4}$, we get, $1.15m + 4\ln(n).$

so we can see we made progress. Roughly, of the form $(1+\epsilon)*m + \epsilon^{-1}*\ln(n)$.

Uses of Randomized Weighted Majority Algorithm (RWMA)

The Randomized Weighted Majority Algorithm can be used to combine multiple algorithms in which case RWMA can be expected to perform nearly as well as the best of the original algorithms in hindsight.

Furthermore, one can apply the Randomized Weighted Majority Algorithm in situations where experts are making choices that cannot be combined (or can't be combined easily). For example, RWMA can be applied to repeated game-playing or the online shortest path problem. In the online shortest path problem, each expert is telling you a different way to drive to work. You pick one path using RWMA. Later you find out how well you would have done using all of the suggested paths and penalize appropriately. To do this right, we want to generalize from "losses" of 0 or 1 to losses in [0,1]. The goal is to have an expected loss not much larger than the loss of the best expert. We can generalize the RWMA by applying a penalty of β^{loss} (i.e. two losses of one half result in the same weight as one loss of 1 and one loss of 0).

Extensions

- Multi-armed bandit problem.

- Efficient algorithm for some cases with many experts.

- Sleeping experts/"specialists" setting.

k-nearest Neighbors Algorithm

In pattern recognition, the *k*-nearest neighbors algorithm (*k*-NN) is a non-parametric method used for classification and regression. In both cases, the input consists of the *k* closest training examples in the feature space. The output depends on whether *k*-NN is used for classification or regression:

- In *k-NN classification*, the output is a class membership. An object is classified by a majority vote of its neighbors, with the object being assigned to the class most common among its *k* nearest neighbors (*k* is a positive integer, typically small). If *k* = 1, then the object is simply assigned to the class of that single nearest neighbor.

- In *k-NN regression*, the output is the property value for the object. This value is the average of the values of its *k* nearest neighbors.

k-NN is a type of instance-based learning, or lazy learning, where the function is only approximated locally and all computation is deferred until classification. The *k*-NN algorithm is among the simplest of all machine learning algorithms.

Both for classification and regression, it can be useful to assign weight to the contributions of the neighbors, so that the nearer neighbors contribute more to the average than the more distant ones. For example, a common weighting scheme consists in giving each neighbor a weight of $1/d$, where d is the distance to the neighbor.

The neighbors are taken from a set of objects for which the class (for *k*-NN classification) or the object property value (for *k*-NN regression) is known. This can be thought of as the training set for the algorithm, though no explicit training step is required.

A peculiarity of the *k*-NN algorithm is that it is sensitive to the local structure of the data. The algorithm is different than *k*-means, another popular machine learning technique.

Statistical Setting

Suppose we have pairs $(X, Y), (X_1, Y_1), \ldots, (X_n, Y_n)$ taking values in $\mathbb{R}^d \times \{1, 2\}$, where Y is the class label of X, so that $X \mid Y = r \sim P_r$ for $r = 1, 2$ (and probability distributions P_r). Given some norm $\|\cdot\|$ on $\mathbb{R}^d$ and a point $x \in \mathbb{R}^d$, let $(X_{(1)}, Y_{(1)}), \ldots, (X_{(n)}, Y_{(n)})$ be a reordering of the training data such that $\| X_{(1)} - x \| \leq \ldots \leq \| X_{(n)} - x \|$.

Algorithm

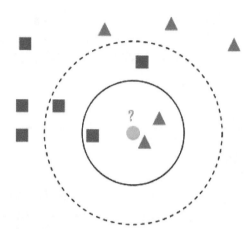

Example of *k*-NN classification. The test sample (green circle) should be classified either to the first class of blue squares or to the second class of red triangles. If $k = 3$ (solid line circle) it is assigned to the second class because there are 2 triangles and only 1 square inside the inner circle. If $k = 5$ (dashed line circle) it is assigned to the first class (3 squares vs. 2 triangles inside the outer circle).

The training examples are vectors in a multidimensional feature space, each with a class label. The training phase of the algorithm consists only of storing the feature vectors and class labels of the training samples.

In the classification phase, k is a user-defined constant, and an unlabeled vector (a query or test point) is classified by assigning the label which is most frequent among the k training samples nearest to that query point.

A commonly used distance metric for continuous variables is Euclidean distance. For discrete variables, such as for text classification, another metric can be used, such as the overlap metric (or Hamming distance). In the context of gene expression microarray data, for example, *k*-NN has also been employed with correlation coefficients such as Pearson and Spearman. Often, the classification accuracy of *k*-NN can be improved significantly if the distance metric is learned with specialized algorithms such as Large Margin Nearest Neighbor or Neighbourhood components analysis.

A drawback of the basic "majority voting" classification occurs when the class distribution is skewed. That is, examples of a more frequent class tend to dominate the prediction of the new example, because they tend to be common among the k nearest neighbors due to their large number. One way to overcome this problem is to weight the classification, taking into account the distance from the test point to each of its k nearest neighbors. The class (or value, in regression problems) of each of the k nearest points is multiplied by a weight proportional to the inverse of the distance from that point to the test point. Another way to overcome skew is by abstraction in data representation. For example, in a self-organizing map (SOM), each node is a representative (a center) of a cluster of similar points, regardless of their density in the original training data. *K*-NN can then be applied to the SOM.

Parameter Selection

The best choice of k depends upon the data; generally, larger values of k reduce the effect of noise on the classification, but make boundaries between classes less distinct. A good k can be selected by various heuristic techniques. The special case where the class is predicted to be the class of the closest training sample (i.e. when $k = 1$) is called the nearest neighbor algorithm.

The accuracy of the k-NN algorithm can be severely degraded by the presence of noisy or irrelevant features, or if the feature scales are not consistent with their importance. Much research effort has been put into selecting or scaling features to improve classification. A particularly popular approach is the use of evolutionary algorithms to optimize feature scaling. Another popular approach is to scale features by the mutual information of the training data with the training classes.

In binary (two class) classification problems, it is helpful to choose k to be an odd number as this avoids tied votes. One popular way of choosing the empirically optimal k in this setting is via bootstrap method.

The 1-nearest Neighbour Classifier

The most intuitive nearest neighbour type classifier is the one nearest neighbour classifier that assigns a point x to the class of its closest neighbour in the feature space, that is $C_n^{1nn}(x) = Y_{(1)}$.

As the size of training data set approaches infinity, the one nearest neighbour classifier guarantees an error rate of no worse than twice the Bayes error rate (the minimum achievable error rate given the distribution of the data).

The Weighted Nearest Neighbour Classifier

The k-nearest neighbour classifier can be viewed as assigning the k nearest neighbours a weight $1/k$ and all others o weight. This can be generalised to weighted nearest neighbour classifiers. That is, where the ith nearest neighbour is assigned a weight w_{ni}, with $\sum_{i=1}^{n} w_{ni} = 1$. An analogous result on the strong consistency of weighted nearest neighbour classifiers also holds.

Let C_n^{wnn} denote the weighted nearest classifier with weights $\{w_{ni}\}_{i=1}^{n}$. Subject to regularity conditions on to class distributions the excess risk has the following asymptotic expansion

$$\mathcal{R}_\mathcal{R}(C_n^{wnn}) - \mathcal{R}_\mathcal{R}(C^{Bayes}) = \left(B_1 s_n^2 + B_2 t_n^2\right)\{1 + o(1)\}$$

for constants B_1 and B_2 where $s_n^2 = \sum_{i=1}^{n} w_{ni}^2$ and $t_n = n^{-2/d} \sum_{i=1}^{n} w_{ni} \{i^{1+2/d} - (i-1)^{1+2/d}\}$.

The optimal weighting scheme $\{w_{ni}^*\}_{i=1}^{n}$, that balances the two terms in the display above, is given as follows: set $k^* = \lfloor Bn^{\frac{4}{d+4}} \rfloor$,

$$w_{ni}^* = \frac{1}{k^*}\left[1 + \frac{d}{2} - \frac{d}{2k^{*2/d}}\{i^{1+2/d} - (i-1)^{1+2/d}\}\right] \text{ for } i = 1, 2, \ldots, k^* \text{ and}$$

$w_{ni}^* = 0$ for $i = k^* + 1, \ldots, n$.

With optimal weights the dominant term in the asymptotic expansion of the excess risk is $\mathcal{O}(n^{-\frac{4}{d+4}})$. Similar results are true when using a bagged nearest neighbour classifier.

Properties

k-NN is a special case of a variable-bandwidth, kernel density "balloon" estimator with a uniform kernel.

The naive version of the algorithm is easy to implement by computing the distances from the test example to all stored examples, but it is computationally intensive for large training sets. Using an appropriate nearest neighbor search algorithm makes k-NN computationally tractable even for large data sets. Many nearest neighbor search algorithms have been proposed over the years; these generally seek to reduce the number of distance evaluations actually performed.

k-NN has some strong consistency results. As the amount of data approaches infinity, the two-class k-NN algorithm is guaranteed to yield an error rate no worse than twice the Bayes error rate (the minimum achievable error rate given the distribution of the data). Various improvements to the k-NN speed are possible by using proximity graphs.

For multi-class k-NN classification, Cover and Hart (1967) prove an upper bound error rate of

$$R^* \le R_{kNN} \le R^*(2 - MR^* / (M - 1))$$

where R^* is the Bayes error rate (which is the minimal error rate possible), R_{kNN} is the k-NN error rate, and M is the number of classes in the problem. For $M = 2$ and as the Bayesian error rate R^* approaches zero, this limit reduces to "not more than twice the Bayesian error rate".

Error Rates

There are many results on the error rate of the k nearest neighbour classifiers. The k-nearest neighbour classifier is strongly (that is for any joint distribution on (X, Y)) consistent provided $k := k_n$ diverges and k_n / n converges to zero as $n \to \infty$.

Let C_n^{knn} denote the k nearest neighbour classifier based on a training set of size n. Under certain regularity conditions, the excess risk yields the following asymptotic expansion

$$\mathcal{R}_{\mathcal{R}}(C_n^{knn}) - \mathcal{R}_{\mathcal{R}}(C^{Bayes}) = \left\{ B_1 \frac{1}{k} + B_2 \left(\frac{k}{n} \right)^{4/d} \right\} \{1 + o(1)\},$$

for some constants B_1 and B_2.

The choice $k^* = \lfloor B n^{\frac{4}{d+4}} \rfloor$ offers a trade off between the two terms in the above display, for which the k^*-nearest neighbour error converges to the Bayes error at the optimal (minimax) rate $\mathcal{O}(n^{-\frac{4}{d+4}})$.

Metric Learning

The K-nearest neighbor classification performance can often be significantly improved through (supervised) metric learning. Popular algorithms are neighbourhood components analysis and large margin nearest neighbor. Supervised metric learning algorithms use the label information to learn a new metric or pseudo-metric.

Feature Extraction

When the input data to an algorithm is too large to be processed and it is suspected to be redundant (e.g. the same measurement in both feet and meters) then the input data will be transformed into a reduced representation set of features (also named features vector). Transforming the input data into the set of features is called feature extraction. If the features extracted are carefully chosen it is expected that the features set will extract the relevant information from the input data in order to perform the desired task using this reduced representation instead of the full size input. Feature extraction is performed on raw data prior to applying k-NN algorithm on the transformed data in feature space.

An example of a typical computer vision computation pipeline for face recognition using k-NN including feature extraction and dimension reduction pre-processing steps (usually implemented with OpenCV):

1. Haar face detection

2. Mean-shift tracking analysis

3. PCA or Fisher LDA projection into feature space, followed by k-NN classification

Dimension Reduction

For high-dimensional data (e.g., with number of dimensions more than 10) dimension reduction is usually performed prior to applying the k-NN algorithm in order to avoid the effects of the curse of dimensionality.

The curse of dimensionality in the k-NN context basically means that Euclidean distance is unhelpful in high dimensions because all vectors are almost equidistant to the search query vector (imagine multiple points lying more or less on a circle with the query point at the center; the distance from the query to all data points in the search space is almost the same).

Feature extraction and dimension reduction can be combined in one step using principal component analysis (PCA), linear discriminant analysis (LDA), or canonical correlation analysis (CCA) techniques as a pre-processing step, followed by clustering by k-NN on feature vectors in reduced-dimension space. In machine learning this process is also called low-dimensional embedding.

For very-high-dimensional datasets (e.g. when performing a similarity search on live video streams, DNA data or high-dimensional time series) running a fast approximate k-NN search using locality sensitive hashing, "random projections", "sketches" or other high-dimensional similarity search techniques from the VLDB toolbox might be the only feasible option.

Decision Boundary

Nearest neighbor rules in effect implicitly compute the decision boundary. It is also possible to compute the decision boundary explicitly, and to do so efficiently, so that the computational complexity is a function of the boundary complexity.

Data Reduction

Data reduction is one of the most important problems for work with huge data sets. Usually, only some of the data points are needed for accurate classification. Those data are called the *prototypes* and can be found as follows:

1. Select the *class-outliers*, that is, training data that are classified incorrectly by k-NN (for a given k)

2. Separate the rest of the data into two sets: (i) the prototypes that are used for the classification decisions and (ii) the *absorbed points* that can be correctly classified by k-NN using prototypes. The absorbed points can then be removed from the training set.

Selection of Class-outliers

A training example surrounded by examples of other classes is called a class outlier. Causes of class outliers include:

*   random error

*   insufficient training examples of this class (an isolated example appears instead of a cluster)

*   missing important features (the classes are separated in other dimensions which we do not know)

*   too many training examples of other classes (unbalanced classes) that create a "hostile" background for the given small class

Class outliers with k-NN produce noise. They can be detected and separated for future analysis. Given two natural numbers, $k>r>0$, a training example is called a (k,r)NN class-outlier if its k nearest neighbors include more than r examples of other classes.

CNN for Data Reduction

Condensed nearest neighbor (CNN, the *Hart algorithm*) is an algorithm designed to reduce the data set for k-NN classification. It selects the set of prototypes U from the training data, such that 1NN with U can classify the examples almost as accurately as 1NN does with the whole data set.

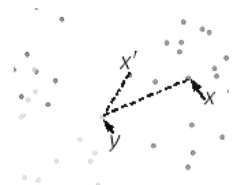

Calculation of the border ratio.

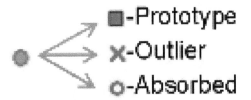

Three types of points: prototypes, class-outliers, and absorbed points.

Given a training set X, CNN works iteratively:

1. Scan all elements of X, looking for an element x whose nearest prototype from U has a different label than x.

2. Remove x from X and add it to U

3. Repeat the scan until no more prototypes are added to U.

Use U instead of X for classification. The examples that are not prototypes are called "absorbed" points.

It is efficient to scan the training examples in order of decreasing border ratio. The border ratio of a training example x is defined as

$$a(x) = \frac{\|x'-y\|}{\|x-y\|}$$

where $\|x\text{-}y\|$ is the distance to the closest example y having a different color than x, and $\|x'\text{-}y\|$ is the distance from y to its closest example x' with the same label as x.

The border ratio is in the interval [0,1] because $\|x'\text{-}y\|$ never exceeds $\|x\text{-}y\|$. This ordering gives preference to the borders of the classes for inclusion in the set of prototypes U. A point of a different label than x is called external to x. The calculation of the border ratio is illustrated by the figure. The data points are labeled by colors: the initial point is x and its label is red. External points are blue and green. The closest to x external point is y. The closest to y red point is x'. The border ratio $a(x) = \|x'\text{-}y\| / \|x\text{-}y\|$ is the attribute of the initial point x.

Below is an illustration of CNN in a series of figures. There are three classes (red, green and blue). Fig. a initially there are 60 points in each class. Fig. b shows the 1NN classification map: each pixel is classified by 1NN using all the data. Fig. c shows the 5NN classification map. White areas correspond to the unclassified regions, where 5NN voting is tied (for example, if there are two green, two red and one blue points among 5 nearest neighbors). Fig. d shows the reduced data set. The crosses are the class-outliers selected by the (3,2)NN rule (all the three nearest neighbors of these instances belong to other classes); the squares are the prototypes, and the empty circles are the absorbed points. The left bottom corner shows the numbers of the class-outliers, prototypes and absorbed points for all three classes. The number of prototypes varies from 15% to 20% for different classes in this example. Fig. e shows that the 1NN classification map with the prototypes is very similar to that with the initial data set. The figures were produced using the Mirkes applet.

- CNN model reduction for k-NN classifiers

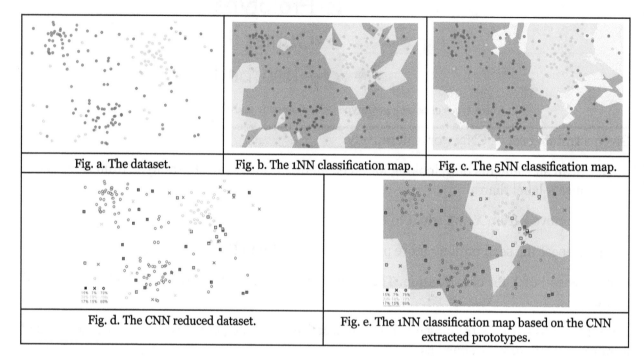

Fig. a. The dataset.	Fig. b. The 1NN classification map.	Fig. c. The 5NN classification map.
Fig. d. The CNN reduced dataset.		Fig. e. The 1NN classification map based on the CNN extracted prototypes.

k-NN Regression

In *k*-NN regression, the *k*-NN algorithm is used for estimating continuous variables. One such algorithm uses a weighted average of the *k* nearest neighbors, weighted by the inverse of their distance. This algorithm works as follows:

1. Compute the Euclidean or Mahalanobis distance from the query example to the labeled examples.

2. Order the labeled examples by increasing distance.

3. Find a heuristically optimal number *k* of nearest neighbors, based on RMSE. This is done using cross validation.

4. Calculate an inverse distance weighted average with the *k*-nearest multivariate neighbors.

k-NN outlier

The distance to the *k*th nearest neighbor can also be seen as a local density estimate and thus is also a popular outlier score in anomaly detection. The larger the distance to the *k*-NN, the lower the local density, the more likely the query point is an outlier. Although quite simple, this outlier model, along with another classic data mining method, local outlier factor, works quite well also in comparison to more recent and more complex approaches, according to a large scale experimental analysis.

Validation of Results

A confusion matrix or "matching matrix" is often used as a tool to validate the accuracy of *k*-NN classification. More robust statistical methods such as likelihood-ratio test can also be applied.

Boosting (Machine Learning)

Boosting is a machine learning ensemble meta-algorithm for primarily reducing bias, and also variance in supervised learning, and a family of machine learning algorithms which convert weak learners to strong ones. Boosting is based on the question posed by Kearns and Valiant (1988, 1989): Can a set of weak learners create a single strong learner? A weak learner is defined to be a classifier which is only slightly correlated with the true classification (it can label examples better than random guessing). In contrast, a strong learner is a classifier that is arbitrarily well-correlated with the true classification.

Robert Schapire's affirmative answer in a 1990 paper to the question of Kearns and Valiant has had significant ramifications in machine learning and statistics, most notably leading to the development of boosting.

When first introduced, the *hypothesis boosting problem* simply referred to the process of turning a weak learner into a strong learner. "Informally, [the hypothesis boosting] problem asks whether an efficient learning algorithm [...] that outputs a hypothesis whose performance is only slightly better than random guessing [i.e. a weak learner] implies the existence of an efficient algorithm that outputs a hypothesis of arbitrary accuracy [i.e. a strong learner]." Algorithms that achieve hypothesis boosting quickly became simply known as "boosting". Freund and Schapire's arcing (Adapt[at]ive Resampling and Combining), as a general technique, is more or less synonymous with boosting.

Boosting Algorithms

While boosting is not algorithmically constrained, most boosting algorithms consist of iteratively learning weak classifiers with respect to a distribution and adding them to a final strong classifier. When they are added, they are typically weighted in some way that is usually related to the weak learners' accuracy. After a weak learner is added, the data are reweighted: examples that are misclassified gain weight and examples that are classified correctly lose weight (some boosting algorithms actually decrease the weight of repeatedly misclassified examples, e.g., boost by majority and BrownBoost). Thus, future weak learners focus more on the examples that previous weak learners misclassified.

There are many boosting algorithms. The original ones, proposed by Robert Schapire (a recursive majority gate formulation) and Yoav Freund (boost by majority), were not adaptive and could not take full advantage of the weak learners. However, Schapire and Freund then developed AdaBoost, an adaptive boosting algorithm that won the prestigious Gödel Prize.

Only algorithms that are provable boosting algorithms in the probably approximately correct learning formulation can accurately be called *boosting algorithms*. Other algorithms that are similar in spirit to boosting algorithms are sometimes called "leveraging algorithms", although they are also sometimes incorrectly called boosting algorithms.

The main variation between many boosting algorithms is their method of weighting training data points and hypotheses. AdaBoost is very popular and perhaps the most significant historically as it was the first algorithm that could adapt to the weak learners. However, there are many more recent algorithms such as LPBoost, TotalBoost, BrownBoost, xgboost, MadaBoost, LogitBoost, and others. Many boosting algorithms fit into the AnyBoost framework, which shows that boosting performs gradient descent in function space using a convex cost function.

Object Categorization

Given images containing various known objects in the world, a classifier can be learned from them to automatically categorize the objects in future images. Simple classifiers built based on some image feature of the object tend to be weak in categorization performance. Using boosting methods for object categorization is a way to unify the weak classifiers in a special way to boost the overall ability of categorization.

Problem of Object Categorization

Object categorization is a typical task of computer vision which involves determining whether or not an image contains some specific category of object. The idea is closely related with recognition, identification, and detection. Appearance based object categorization typically contains feature extraction, learning a classifier, and applying the classifier to new examples. There are many ways to represent a category of objects, e.g. from shape analysis, bag of words models, or local descriptors such as SIFT, etc. Examples of supervised classifiers are Naive Bayes classifier, SVM, mixtures of Gaussians, neural network, etc. However, research has shown that object categories and their locations in images can be discovered in an unsupervised manner as well.

Status Quo for Object Categorization

The recognition of object categories in images is a challenging problem in computer vision, especially when the number of categories is large. This is due to high intra class variability and the need for generalization across variations of objects within the same category. Objects within one category may look quite different. Even the same object may appear unalike under different viewpoint, scale, and illumination. Background clutter and partial occlusion add difficulties to recognition as well. Humans are able to recognize thousands of object types, whereas most of the existing object recognition systems are trained to recognize only a few, e.g., human face, car, simple objects, etc. Research has been very active on dealing with more categories and enabling incremental additions of new categories, and although the general problem remains unsolved, several multi-category objects detectors (number of categories around 20) for clustered scenes have been developed. One means is by feature sharing and boosting.

Boosting for Binary Categorization

AdaBoost can be used for face detection as an example of binary categorization. The two categories are faces versus background. The general algorithm is as follows:

1. Form a large set of simple features

2. Initialize weights for training images

3. For T rounds

1. Normalize the weights

2. For available features from the set, train a classifier using a single feature and evaluate the training error

3. Choose the classifier with the lowest error

4. Update the weights of the training images: increase if classified wrongly by this classifier, decrease if correctly

4. Form the final strong classifier as the linear combination of the T classifiers (coefficient larger if training error is small)

After boosting, a classifier constructed from 200 features could yield a 95% detection rate under a 10^{-5} false positive rate·

Another application of boosting for binary categorization is a system which detects pedestrians using patterns of motion and appearance. This work is the first to combine both motion information and appearance information as features to detect a walking person. It takes a similar approach as the face detection work of Viola and Jones.

Boosting for Multi-class Categorization

Compared with binary categorization, multi-class categorization looks for common features that can be shared across the categories at the same time. They turn to be more generic edge like features. During learning, the detectors for each category can be trained jointly. Compared with training separately, it generalizes better, needs less training data, and requires less number of features to achieve same performance.

The main flow of the algorithm is similar to the binary case. What is different is that a measure of the joint training error shall be defined in advance. During each iteration the algorithm chooses a classifier of a single feature (features which can be shared by more categories shall be encouraged). This can be done via converting multi-class classification into a binary one (a set of categories versus the rest), or by introducing a penalty error from the categories which do not have the feature of the classifier.

In the paper "Sharing visual features for multiclass and multiview object detection", A. Torralba et al. used GentleBoost for boosting and showed that when training data is limited, learning via sharing features does a much better job than no sharing, given same boosting rounds. Also, for a given performance level, the total number of features required (and therefore the run time cost of the classifier) for the feature sharing detectors, is observed to scale approximately logarithmically with the number of class, i.e., slower than linear growth in the non-sharing case. Similar results are shown in the paper "Incremental learning of object detectors using a visual shape alphabet", yet the authors used AdaBoost for boosting.

Criticism

In 2008 Phillip Long (at Google) and Rocco A. Servedio (Columbia University) published a paper at the 25th International Conference for Machine Learning suggesting that many of these algorithms are probably flawed. They conclude that "convex potential boosters cannot withstand random classification noise," thus making the applicability of such algorithms for real world, noisy data sets questionable. The paper shows that if any fraction of the training data is mis-labeled, the boosting algorithm tries extremely hard to correctly classify these training examples, and fails to produce a model with accuracy better than 1/2. This result does not apply to branching program based boosters but does apply to AdaBoost, LogitBoost, and others.

Bootstrap Aggregating

Bootstrap aggregating, also called bagging, is a machine learning ensemble meta-algorithm designed to improve the stability and accuracy of machine learning algorithms used in statistical classification and regression. It also reduces variance and helps to avoid overfitting. Although it is usually applied to decision tree methods, it can be used with any type of method. Bagging is a special case of the model averaging approach.

History

Bagging (Bootstrap aggregating) was proposed by Leo Breiman in 1994 to improve the classification by combining classifications of randomly generated training sets.1994. Technical Report No. 421.

Description of the Technique

Given a standard training set D of size n, bagging generates m new training sets D_i, each of size n', by sampling from D uniformly and with replacement. By sampling with replacement, some observations may be repeated in each D_i. If $n'=n$, then for large n the set D_i is expected to have the fraction $(1 - 1/e)$ ($\approx$63.2%) of the unique examples of D, the rest being duplicates. This kind of sample is known as a bootstrap sample. The m models are fitted using the above m bootstrap samples and combined by averaging the output (for regression) or voting (for classification).

Bagging leads to "improvements for unstable procedures" (Breiman, 1996), which include, for example, artificial neural networks, classification and regression trees, and subset selection in linear regression (Breiman, 1994). An interesting application of bagging showing improvement in preimage learning is provided here. On the other hand, it can mildly degrade the performance of stable methods such as K-nearest neighbors (Breiman, 1996).

Example: Ozone Data

To illustrate the basic principles of bagging, below is an analysis on the relationship between ozone and temperature (data from Rousseeuw and Leroy (1986), analysis done in R).

The relationship between temperature and ozone in this data set is apparently non-linear, based on the scatter plot. To mathematically describe this relationship, LOESS smoothers (with span 0.5) are used. Instead of building a single smoother from the complete data set, 100 bootstrap samples of the data were drawn. Each sample is different from the original data set, yet resembles it in distribution and variability. For each bootstrap sample, a LOESS smoother was fit. Predictions from these 100 smoothers were then made across the range of the data. The first 10 predicted smooth fits appear as grey lines in the figure below. The lines are clearly very *wiggly* and they overfit the data - a result of the span being too low.

By taking the average of 100 smoothers, each fitted to a subset of the original data set, we arrive at one bagged predictor (red line). Clearly, the mean is more stable and there is less overfit.

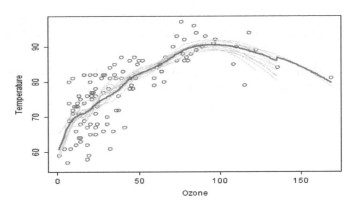

References

- Dempster, A.P.; Laird, N.M.; Rubin, D.B. (1977). "Maximum Likelihood from Incomplete Data via the EM Algorithm". Journal of the Royal Statistical Society, Series B. 39 (1): 1–38. JSTOR 2984875. MR 0501537

- Stuart Russell and Peter Norvig (2010). Artificial Intelligence A Modern Approach 3rd Edition. Upper Saddle River, New Jersey: Pearson Education/Prentice-Hall. ISBN 978-0-13-604259-4

- Ikeda, Shiro; Amari, Shun-ichi; Nakahara, Hiroyuki. "Convergence of The Wake-Sleep Algorithm" (PDF). The Institute of Statistical Mathematics. Retrieved 2015-11-01

- Hotelling, H (1936). "Relations between two sets of variates" (PDF). Biometrika. 28 (3-4): 321. doi:10.2307/2333955

- Sundberg, Rolf (1974). "Maximum likelihood theory for incomplete data from an exponential family". Scandinavian Journal of Statistics. 1 (2): 49–58. JSTOR 4615553. MR 381110

- Ramaswamy, S.; Rastogi, R.; Shim, K. (2000). Efficient algorithms for mining outliers from large data sets. Proceedings of the 2000 ACM SIGMOD international conference on Management of data – SIGMOD '00. p. 427. ISBN 1-58113-217-4. doi:10.1145/342009.335437

- Maei, Hamid Reza (2007-01-25). "Wake-sleep algorithm for representational learning". University of Montreal. Retrieved 2011-11-01

- Belkin, M; P Niyogi (2003). "Laplacian eigenmaps for dimensionality reduction and data representation" (PDF). Neural Computation. 15: 1373–1396. doi:10.1162/089976603321780317

- Wolynetz, M.S. (1979). "Maximum likelihood estimation in a linear model from confined and censored normal data". Journal of the Royal Statistical Society, Series C. 28 (2): 195–206. doi:10.2307/2346749

- Little, Roderick J.A.; Rubin, Donald B. (1987). Statistical Analysis with Missing Data. Wiley Series in Probability and Mathematical Statistics. New York: John Wiley & Sons. pp. 134–136. ISBN 0-471-80254-9

- Littlestone, N.; Warmuth, M. (1994). "The Weighted Majority Algorithm". Information and Computation. 108: 212–261. doi:10.1006/inco.1994.1009

- Jamshidian, Mortaza; Jennrich, Robert I. (1997). "Acceleration of the EM Algorithm by using Quasi-Newton Methods". Journal of the Royal Statistical Society, Series B. 59 (2): 569–587. MR 1452026. doi:10.1111/1467-9868.00083

- Hastie, Trevor; Tibshirani, Robert; Friedman, Jerome (2001). "8.5 The EM algorithm". The Elements of Statistical Learning. New York: Springer. pp. 236–243. ISBN 0-387-95284-5

- Hinton, Geoffrey; Dayan, Peter; Frey, Brendan J; Neal, Radford M (1995-04-03). "The wake-sleep algorithm for unsupervised neural networks" (PDF). Retrieved 2015-11-01

- Cover TM, Hart PE (1967). "Nearest neighbor pattern classification". IEEE Transactions on Information Theory. 13 (1): 21–27. doi:10.1109/TIT.1967.1053964

- Toussaint GT (April 2005). "Geometric proximity graphs for improving nearest neighbor methods in instance-based learning and data mining". International Journal of Computational Geometry and Applications. 15 (2): 101–150. doi:10.1142/S0218195905001622

- Altman, N. S. (1992). "An introduction to kernel and nearest-neighbor nonparametric regression". The American Statistician. 46 (3): 175–185. doi:10.1080/00031305.1992.10475879

Allied Fields of Machine Learning

Meta leaning is a branch of machine learning which applies automatic learning algorithms to meta-data. This process is used to understand how automatic leaning can be made more flexible and can be used in solving difficult learning problems. This chapter is a compilation of the various allied fields of machine learning that form an integral part of the broader subject matter.

Meta Learning (Computer Science)

Meta learning is a subfield of Machine learning where automatic learning algorithms are applied on meta-data about machine learning experiments. Although different researchers hold different views as to what the term exactly means, the main goal is to use such meta-data to understand how automatic learning can become flexible in solving different kinds of learning problems, hence to improve the performance of existing learning algorithms.

Flexibility is very important because each learning algorithm is based on a set of assumptions about the data, its inductive bias. This means that it will only learn well if the bias matches the data in the learning problem. A learning algorithm may perform very well on one learning problem, but very badly on the next. From a non-expert point of view, this poses strong restrictions on the use of machine learning or data mining techniques, since the relationship between the learning problem (often some kind of database) and the effectiveness of different learning algorithms is not yet understood.

By using different kinds of meta-data, like properties of the learning problem, algorithm properties (like performance measures), or patterns previously derived from the data, it is possible to select, alter or combine different learning algorithms to effectively solve a given learning problem. Critiques of meta learning approaches bear a strong resemblance to the critique of metaheuristic, which can be said to be a related problem.

Definition

A proposed definition for what qualifies as a meta learning system considers three requirements:

1. The system must include a learning subsystem, which adapts with experience.

2. Experience is gained by exploiting meta knowledge extracted

   o ...in a previous learning episode on a single dataset.

   o ...from different domains or problems.

3. Learning bias must be chosen dynamically.

The term bias in the last point refers to the set of assumptions influencing the choice of hypotheses for explaining the data and must not be confused with the notion of bias represented in the bias-variance dilemma. Meta learning is concerned with two aspects of learning bias; declarative bias specifies the representation of the space of hypotheses, and affects the size of the search space (i.e. represent hypotheses using linear functions only) while procedural bias imposes constraints on the ordering of the inductive hypotheses (i.e. preferring smaller hypotheses).

Different Views on Meta Learning

These are some of the views on (and approaches to) meta learning, please note that there exist many variations on these general approaches:

- *Discovering meta-knowledge* works by inducing knowledge (e.g. rules) that expresses how each learning method will perform on different learning problems. The meta-data is formed by characteristics of the data (general, statistical, information-theoretic etc.) in the learning problem, and characteristics of the learning algorithm (type, parameter settings, performance measures,...). Another learning algorithm then learns how the data characteristics relate to the algorithm characteristics. Given a new learning problem, the data characteristics are measured, and the performance of different learning algorithms can be predicted. Hence, one can select the algorithms best suited for the new problem, at least if the induced relationship holds.

- *Stacked generalisation* works by combining a number of (different) learning algorithms. The meta-data is formed by the predictions of those different algorithms. Then another learning algorithm learns from this meta-data to predict which combinations of algorithms give generally good results. Given a new learning problem, the predictions of the selected set of algorithms are combined (e.g. by (weighted) voting) to provide the final prediction. Since each algorithm is deemed to work on a subset of problems, a combination is hoped to be more flexible and still able to make good predictions.

- *Boosting* is related to stacked generalisation, but uses the same algorithm multiple times, where the examples in the training data get different weights over each run. This yields different predictions, each focused on rightly predicting a subset of the data, and combining those predictions leads to better (but more expensive) results.

- *Dynamic bias selection* works by altering the inductive bias of a learning algorithm to match the given problem. This is done by altering key aspects of the learning algorithm, such as the hypothesis representation, heuristic formulae, or parameters. Many different approaches exist.

- *Inductive transfer* also called learning to learn, studies how the learning process can be improved over time. Meta-data consists of knowledge about previous learning episodes, and is used to efficiently develop an effective hypothesis for a new task. A related approach is called learning to learn, in which the goal is to use acquired knowledge from one domain to help learning in other domains.

- Other approaches using meta-data to improve automatic learning are learning classifier systems, case-based reasoning and constraint satisfaction.

Adversarial Machine Learning

Adversarial machine learning is a research field that lies at the intersection of machine learning and computer security. It aims to enable the safe adoption of machine learning techniques in adversarial settings like spam filtering, malware detection and biometric recognition.

The problem arises from the fact that machine learning techniques were originally designed for stationary environments in which the training and test data are assumed to be generated from the same (although possibly unknown) distribution. In the presence of intelligent and adaptive adversaries, however, this working hypothesis is likely to be violated to at least some degree (depending on the adversary). In fact, a malicious adversary can carefully manipulate the input data exploiting specific vulnerabilities of learning algorithms to compromise the whole system security.

Examples include: attacks in spam filtering, where spam messages are obfuscated through misspelling of bad words or insertion of good words; attacks in computer security, e.g., to obfuscate malware code within network packets or mislead signature detection; attacks in biometric recognition, where fake biometric traits may be exploited to impersonate a legitimate user (biometric spoofing) or to compromise users' template galleries that are adaptively updated over time.

Security Evaluation

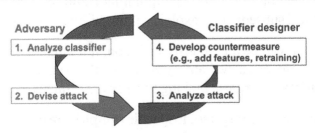

Conceptual representation of the reactive arms race.

To understand the security properties of learning algorithms in adversarial settings, one should address the following main issues:

- identifying potential vulnerabilities of machine learning algorithms during learning and classification;

- devising appropriate attacks that correspond to the identified threats and evaluating their impact on the targeted system;

- proposing countermeasures to improve the security of machine learning algorithms against the considered attacks.

This process amounts to simulating a proactive arms race (instead of a reactive one, as depicted in figures, where system designers try to anticipate the adversary in order to understand whether there are potential vulnerabilities that should be fixed in advance; for instance, by means of specific countermeasures such as additional features or different learning algorithms. However proactive approaches are not necessarily superior to reactive ones. For instance, in, the authors showed that under some circumstances, reactive approaches are more suitable for improving system security.

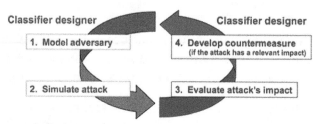

Conceptual representation of the proactive arms race.

Attacks Against Machine Learning Algorithms (Supervised)

The first step of the above-sketched arms race is identifying potential attacks against machine learning algorithms. A substantial amount of work has been done in this direction.

A Taxonomy of Potential Attacks Against Machine Learning

Attacks against (supervised) machine learning algorithms have been categorized along three primary axes: their *influence* on the classifier, the *security violation* they cause, and their *specificity*.

- *Attack influence.* It can be causative, if the attack aims to introduce vulnerabilities (to be exploited at classification phase) by manipulating training data; or exploratory, if the attack aims to find and subsequently exploit vulnerabilities at classification phase.

- *Security violation.* It can be an integrity violation, if it aims to get malicious samples misclassified as legitimate; or an availability violation, if the goal is to increase the misclassification rate of legitimate samples, making the classifier unusable (e.g., a denial of service).

- *Attack specificity.* It can be targeted, if specific samples are considered (e.g., the adversary aims to allow a specific intrusion or she wants a given spam email to get past the filter); or indiscriminate.

This taxonomy has been extended into a more comprehensive threat model that allows one to make explicit assumptions on the adversary's goal, knowledge of the attacked system, capability of manipulating the input data and/or the system components, and on the corresponding (potentially, formally-defined) attack strategy. Details can be found here. Two of the main attack scenarios identified according to this threat model are sketched below.

Evasion Attacks

Evasion attacks are the most prevalent type of attack that may be encountered in adversarial settings during system operation. For instance, spammers and hackers often attempt to evade detection by obfuscating the content of spam emails and malware code. In the evasion setting, malicious samples are modified at test time to evade detection; that is, to be misclassified as legitimate.

No influence over the training data is assumed. A clear example of evasion is image-based spam in which the spam content is embedded within an attached image to evade the textual analysis performed by anti-spam filters. Another example of evasion is given by spoofing attacks against biometric verification systems.

Poisoning Attacks

Machine learning algorithms are often re-trained on data collected during operation to adapt to changes in the underlying data distribution. For instance, intrusion detection systems (IDSs) are often re-trained on a set of samples collected during network operation. Within this scenario, an attacker may poison the training data by injecting carefully designed samples to eventually compromise the whole learning process. Poisoning may thus be regarded as an adversarial contamination of the training data. Examples of poisoning attacks against machine learning algorithms (including learning in the presence of worst-case adversarial label flips in the training data) can be found in.

Attacks Against Clustering Algorithms

Clustering algorithms have been increasingly adopted in security applications to find dangerous or illicit activities. For instance, clustering of malware and computer viruses aims to identify and categorize different existing malware families, and to generate specific signatures for their detection by anti-viruses, or signature-based intrusion detection systems like Snort. However, clustering algorithms have not been originally devised to deal with deliberate attack attempts that are designed to subvert the clustering process itself. Whether clustering can be safely adopted in such settings thus remains questionable. Preliminary work reporting some vulnerability of clustering can be found in.

Secure Learning in Adversarial Settings

A number of defense mechanisms against evasion, poisoning and privacy attacks have been proposed in the field of adversarial machine learning, including:

1. The definition of secure learning algorithms;

2. The use of multiple classifier systems;

3. The use of randomization or disinformation to mislead the attacker while acquiring knowledge of the system;

4. The study of privacy-preserving learning.

5. Ladder algorithm for Kaggle-style competitions.

6. Game theoretic models for adversarial machine learning and data mining.

Software

Some software libraries are available, mainly for testing purposes and research.

- AdversariaLib (includes implementation of evasion attacks from).

- AlfaSVMLib. Adversarial Label Flip Attacks against Support Vector Machines.

- Poisoning Attacks against Support Vector Machines, and Attacks against Clustering Algorithms

- deep-pwning Metasploit for deep learning which currently has attacks on deep neural networks using Tensorflow

Quantum Machine Learning

Quantum machine learning is an emerging interdisciplinary research area at the intersection of quantum physics and machine learning. One can distinguish four different ways of merging the two parent disciplines. Quantum machine learning algorithms can use the advantages of quantum computation in order to improve classical methods of machine learning, for example by developing efficient implementations of expensive classical algorithms on a quantum computer. On the other hand, one can apply classical methods of machine learning to analyse quantum systems. Most generally, one can consider situations wherein both the learning device and the system under study are fully quantum.

A related branch of research explores methodological and structural similarities between certain physical systems and learning systems, in particular neural networks, which has revealed, for example, that certain mathematical and numerical techniques from quantum physics carry over to classical deep learning.

Four different approaches to combine the disciplines of quantum computing and machine learning. The first letter refers to whether the system under study is classical or quantum, while the second letter defines whether a classical or quantum information processing device is used.

Quantum-enhanced Machine Learning

Quantum-enhanced machine learning refers to quantum algorithms that solve tasks in machine learning, thereby improving a classical machine learning method. Such algorithms typically require one to encode the given classical dataset into a quantum computer, so as to make it accessible for quantum information processing. After this, quantum information processing routines can be applied and the result of the quantum computation is read out by measuring the quantum system. For example, the outcome of the measurement of a qubit could reveal the result of a binary classification task. While many proposals of quantum machine learning algorithms are still purely theoretical and require a full-scale universal quantum computer to be tested, others have been implemented on small-scale or special purpose quantum devices.

Linear Algebra Simulation with Quantum Amplitudes

One line of approaches is based on the idea of *amplitude encoding*, that is, to associate the amplitudes of a quantum state with the inputs and outputs of computations. Since a state of n qubits is described by 2^n complex amplitudes, this information encoding can allow for an exponentially compact representation. Intuitively, this corresponds to associating a discrete probability distribution over binary random variables with a classical vector. The goal of algorithms based on amplitude encoding is to formulate quantum algorithms whose resources grow polynomial in the number of qubits n, which amounts to a logarithmic growth in the number of amplitudes and thereby the dimension of the input.

Many quantum machine learning algorithms in this category are based on variations of the quantum algorithm for linear systems of equations which, under specific conditions, performs a matrix inversion using an amount of physical resources growing only logarithmically in the dimensions of the matrix. One of these conditions is that a Hamiltonian which entrywise corresponds to the matrix can be simulated efficiently, which is known to be possible if the matrix is sparse or low rank. For reference, any known classical algorithm for matrix inversion requires a number of operations that grows at least quadratically in the dimension of the matrix.

Quantum matrix inversion can be applied to machine learning methods in which the training reduces to solving a linear system of equations, for example in least-squares linear regression, the least-squares version of support vector machines, and Gaussian processes.

A crucial bottleneck of methods that simulate linear algebra computations with the amplitudes of quantum states is state preparation, which often requires one to initialise a quantum system in a state whose amplitudes reflect the features of the entire dataset. Although efficient methods for state preparation are known for specific cases, this step easily hides the complexity of the task.

Quantum Machine Learning Algorithms Based on Grover Search

Another approach to improving classical machine learning with quantum information processing uses amplitude amplification methods based on Grover's search algorithm, which has been shown to solve unstructured search problems with a quadratic speedup compared to classical algorithms. These quantum routines can be employed for learning algorithms that translate into an unstructured search task, as can be done, for instance, in the case of the k-medians and the k-nearest neighbors algorithms. Another application is a quadratic speedup in the training of perceptron.

Amplitude amplification is often combined with quantum walks to achieve the same quadratic speedup. Quantum walks have been proposed to enhance Google's PageRank algorithm as well as the performance of reinforcement learning agents in the projective simulation framework.

Quantum-enhanced Reinforcement Learning

Reinforcement learning is a third branch of machine learning, distinct from supervised and unsupervised learning, which also admits quantum enhancements. In quantum-enhanced reinforcement learning, a quantum agent interacts with a classical environment and occasionally receives rewards for its actions, which allows the agent to adapt its behaviour—in other words, to learn what to do in order to gain more rewards. In some situations, either because of the quantum processing capability of the agent, or due to the possibility to probe the environment in superpositions, a quantum speedup may be achieved. Implementations of these kinds of protocols in superconducting circuits and in systems of trapped ions have been proposed.

Quantum Sampling Techniques

Sampling from high-dimensional probability distributions is at the core of a wide spectrum of computational techniques with important applications across science, engineering, and society. Examples include deep learning, probabilistic programming, and other machine learning and artificial intelligence applications.

A computationally hard problem, which is key for some relevant machine learning tasks, is the estimation of averages over probabilistic models defined in terms of a Boltzmann distribution. Sampling from generic probabilistic models is hard: algorithms relying heavily on sampling are expected to remain intractable no matter how large and powerful classical computing resources become. Even though quantum annealers, like those produced by D-Wave Systems, were designed for challenging combinatorial optimization problems, it has been recently recognized as a potential candidate to speed up computations that rely on sampling by exploiting quantum effects.

Some research groups have recently explored the use of quantum annealing hardware for training Boltzmann machines and deep neural networks. The standard approach to training Boltzmann machines relies on the computation of certain averages that can be estimated by standard sampling techniques, such as Markov chain Monte Carlo algorithms. Another possibility is to rely on a physical process, like quantum annealing, that naturally generates samples from a Boltzmann distribution. The objective is to find the optimal control parameters that best represent the empirical distribution of a given dataset.

The D-Wave 2X system hosted at NASA Ames Research Center has been recently used for the learning of a special class of restricted Boltzmann machines that can serve as a building block for deep learning architectures. Complementary work that appeared roughly simultaneously showed that quantum annealing can be used for supervised learning in classification tasks. The same device was later used to train a fully connected Boltzmann machine to generate, reconstruct, and classify down-scaled, low-resolution handwritten digits, among other synthetic datasets. In both cases, the models trained by quantum annealing had a similar or better performance in terms of quality. The ultimate question that drives this endeavour is whether there is quantum speedup in sampling applications. Experience with the use of quantum annealers for combinatorial optimization suggests the answer is not straightforward.

Inspired by the success of Boltzmann machines based on classical Boltzmann distribution, a new machine learning approach based on quantum Boltzmann distribution of a transverse-field Ising Hamiltonian was recently proposed. Due to the non-commutative nature of quantum mechanics, the training process of the quantum Boltzmann machine can become nontrivial. This problem was, to some extent, circumvented by introducing bounds on the quantum probabilities, allowing the authors to train the model efficiently by sampling. It is possible that a specific type of quantum

Boltzmann machine has been trained in the D-Wave 2X by using a learning rule analogous to that of classical Boltzmann machines.

Quantum annealing is not the only technology for sampling. In a prepare and measure scenario, a universal quantum computer prepares a thermal state, which is then sampled by measurements. This can reduce the time required to train a deep restricted Boltzmann machine, and provide a richer and more comprehensive framework for deep learning than classical computing. The same quantum methods also permit efficient training of full Boltzmann machines and multi-layer, fully connected models and do not have well-known classical counterparts. Relying on an efficient thermal state preparation protocol starting from an arbitrary state, quantum-enhanced Markov logic networks exploit the symmetries and the locality structure of the probabilistic graphical model generated by a first-order logic template. This provides an exponential reduction in computational complexity in probabilistic inference, and, while the protocol relies on a universal quantum computer, under mild assumptions it can be embedded on contemporary quantum annealing hardware.

Quantum Neural Networks

There are quantum analogues or generalisations of classical neural nets which are known as quantum neural networks.

Quantum Learning Theory

Quantum learning theory pursues a mathematical analysis of the quantum generalizations of classical learning models and of the possible speed-ups or other improvements that they may provide. The framework is very similar to that of classical computational learning theory, but the learner in this case is a quantum information processing device, while the data may be either classical or quantum. Quantum learning theory should be contrasted with the quantum-enhanced machine learning discussed above, where the goal was to consider *specific problems* and to use quantum protocols to improve the time complexity of classical algorithms for these problems. Although quantum learning theory is still under development, partial results in this direction have been obtained.

The starting point in learning theory is typically a *concept class*, a set of possible concepts. Usually a concept is a function on some domain, such as $\{0,1\}^n$. For example, the concept class could be the set of disjunctive normal form (DNF) formulas on n bits or the set of Boolean circuits of some constant depth. The goal for the learner is to learn (exactly or approximately) an unknown *target concept* from this concept class. The learner may be actively interacting with the target concept, or passively receiving samples from it.

In active learning, a learner can make *membership queries* to the target concept c, asking for its value $c(x)$ on inputs x chosen by the learner. The learner then has to reconstruct the exact target concept, with high probability. In the model of *quantum exact learning*, the learner can make membership queries in quantum superposition. If the complexity of the learner is measured by the number of membership queries it makes, then quantum exact learners can be polynomially more efficient than classical learners for some concept classes, but not more. If complexity is measured by the amount of *time* the learner uses, then there are concept classes that can be learned efficiently by quantum learners but not by classical learners (under plausible complexity-theoretic assumptions).

A natural model of passive learning is Valiant's probably approximately correct (PAC) learning. Here the learner receives random examples $(x,c(x))$, where x is distributed according to some unknown distribution D. The learner's goal is to output a hypothesis function h such that $h(x)=c(x)$ with high probability when x is drawn according to D. The learner has to be able to produce such an 'approximately correct' h for every D and every target concept c in its concept class. We can consider replacing the random examples by potentially more powerful quantum examples $\sum_x \sqrt{D(x)} \, | \, x, c(x) \rangle$. In the PAC model (and the related agnostic model), this doesn't significantly reduce the number of examples needed: for every concept class, classical and quantum sample complexity are the same up to constant factors. However, for learning under some fixed distribution D, quantum examples can be very helpful, for example for learning DNF under the uniform distribution. When considering *time* complexity, there exist concept classes that can be PAC-learned efficiently by quantum learners, even from classical examples, but not by classical learners (again, under plausible complexity-theoretic assumptions).

This passive learning type is also the most common scheme in supervised learning: a learning algorithm typically takes the training examples fixed, without the ability to query the label of unlabelled examples. Outputting a hypothesis h is a step of induction. Classically, an inductive model splits into a training and an application phase: the model parameters are estimated in the training phase, and the learned model is applied an arbitrary many times in the application phase. In the asymptotic limit of the number of applications, this splitting of phases is also present with quantum resources.

Classical Learning Applied to Quantum Systems

The term quantum machine learning is also used for approaches that apply classical methods of machine learning to the study of quantum systems. A prime example is the use of classical learning techniques to process large amounts of experimental data in order to characterize an unknown quantum system (for instance in the context of quantum information theory and for the development of quantum technologies), but there are also more exotic applications.

The ability to experimentally control and prepare increasingly complex quantum systems brings with it a growing need to turn large and noisy data sets into meaningful information. This is a problem that has already been studied extensively in the classical setting, and consequently, many existing machine learning techniques can be naturally adapted to more efficiently address experimentally relevant problems. For example, Bayesian methods and concepts of algorithmic learning can be fruitfully applied to tackle quantum state classification, Hamiltonian learning, and the characterization of an unknown unitary transformation. Other problems that have been addressed with this approach are given in the following list:

- Identifying an accurate model for the dynamics of a quantum system, through the reconstruction of the Hamiltonian;

- Extracting information on unknown states;

- Learning unknown unitary transformations and measurements;

- Engineering of quantum gates from qubit networks with pairwise interactions, using time dependent or independent Hamiltonians.

However, the characterization of quantum states and processes is not the only application of classical machine learning techniques. Some additional applications include

- Automatic generation of new quantum experiments;

- Solving the many-body, static and time-dependent Schrödinger equation;

- Identifying phase transitions from entanglement spectra;

- Generating adaptive feedback schemes for quantum metrology.

Fully Quantum Machine Learning

In the most general case of quantum machine learning, both the learning device and the system under study, as well as their interaction, are fully quantum.

One class of problem that can benefit from the fully quantum approach is that of 'learning' unknown quantum states, processes or measurements, in the sense that one can subsequently reproduce them on another quantum system. For example, one may wish to learn a measurement that discriminates between two coherent states, given not a classical description of the states to be discriminated, but instead a set of example quantum systems prepared in these states. The naive approach would be to first extract a classical description of the states and then implement an ideal discriminating measurement based on this information. This would only require classical learning. However, one can show that a fully quantum approach is strictly superior in this case. (This also relates to work on quantum pattern matching.) The problem of learning unitary transformations can be approached in a similar way.

Going beyond the specific problem of learning states and transformations, the task of clustering also admits a fully quantum version, wherein both the oracle which returns the distance between data-points and the information processing device which runs the algorithm are quantum. Finally, a general framework spanning supervised, unsupervised and reinforcement learning in the fully quantum setting was introduced in, where it was also shown that the possibility of probing the environment in superpositions permits a quantum speedup in reinforcement learning.

Implementations and Experiments

The earliest experiments were conducted using the adiabatic D-Wave quantum computer, for instance, to detect cars in digital images using regularized boosting with a nonconvex objective function in a demonstration in 2009. Many experiments followed on the same architecture, and leading tech companies have shown interest in the potential of quantum machine learning for future technological implementations. In 2013, Google Research, NASA, and the Universities Space Research Association launched the Quantum Artificial Intelligence Lab which explores the use of the adiabatic D-Wave quantum computer. A more recent example trained a probabilistic generative models with arbitrary pairwise connectivity, showing that their model is capable of generating handwritten digits as well as reconstructing noisy images of bars and stripes and handwritten digits.

Using a different annealing technology based on nuclear magnetic resonance (NMR), a quantum Hopfield network was implemented in 2009 that mapped the input data and memorized data to

Hamiltonians, allowing the use of adiabatic quantum computation. NMR technology also enables universal quantum computing, and it was used for the first experimental implementation of a quantum support vector machine to distinguish hand written number '6' and '9' on a liquid-state quantum computer in 2015. The training data involved the pre-processing of the image which maps them to normalized 2-dimensional vectors to represent the images as the states of a qubit. The two entries of the vector are the vertical and horizontal ratio of the pixel intensity of the image. Once the vectors are defined on the feature space, the quantum support vector machine was implemented to classify the unknown input vector. The readout avoids costly quantum tomography by reading out the final state in terms of direction (up/down) of the NMR signal.

Photonic implementations are attracting more attention, not the least because they do not require extensive cooling. Simultaneous spoken digit and speaker recognition and chaotic time-series prediction were demonstrated at data rates beyond 1 gigabyte per second in 2013. Using non-linear photonics to implement an all-optical linear classifier, a perceptron model was capable of learning the classification boundary iteratively from training data through a feedback rule. A core building block in many learning algorithms is to calculate the distance between two vectors: this was first to experimentally demonstrated up to eight dimensions using entangled qubits in a photonic quantum computer in 2015.

Recently, based on a neuromimetic approach, a novel ingredient has been added to the field of quantum machine learning, in the form of a so-called quantum memristor, a quantized model of the standard classical memristor. This device can be constructed by means of a tunable resistor, weak measurements on the system, and a classical feed-forward mechanism. An implementation of a quantum memristor in superconducting circuits has been proposed, and an experiment with quantum dots performed. A quantum memristor would implement nonlinear interactions in the quantum dynamics which would aid the search for a fully functional quantum neural network.

Robot Learning

Robot learning is a research field at the intersection of machine learning and robotics. It studies techniques allowing a robot to acquire novel skills or adapt to its environment through learning algorithms. The embodiment of the robot, situated in a physical embedding, provides at the same time specific difficulties (e.g. high-dimensionality, real time constraints for collecting data and learning) and opportunities for guiding the learning process (e.g. sensorimotor synergies, motor primitives).

Example of skills that are targeted by learning algorithms include sensorimotor skills such as locomotion, grasping, active object categorization, as well as interactive skills such as joint manipulation of an object with a human peer, and linguistic skills such as the grounded and situated meaning of human language. Learning can happen either through autonomous self-exploration or through guidance from a human teacher, like for example in robot learning by imitation.

Robot learning can be closely related to adaptive control, reinforcement learning as well as developmental robotics which considers the problem of autonomous lifelong acquisition of repertoires of skills. While machine learning is frequently used by computer vision algorithms employed in the context of robotics, these applications are usually not referred to as "robot learning".

Projects

Maya Cakmak, assistant professor of computer science and engineering at the University of Washington, is trying to create a robot that learns by imitating - a technique called "programming by demonstration". A researcher shows it a cleaning technique for the robot's vision system and it generalizes the cleaning motion from the human demonstration as well as identifying the "state of dirt" before and after cleaning.

Similarly the Baxter industrial robot can be taught how to do something by grabbing its arm and showing it the desired movements. It can also use deep learning to teach itself to grasp an unknown object.

Sharing Learned Skills and Knowledge

In Telex's "Million Object Challenge" the goal is robots that learn how to spot and handle simple items and upload their data to the cloud to allow other robots to analyze and use the information.

RoboBrain is a knowledge engine for robots which can be freely accessed by any device wishing to carry out a task. The database gathers new information about tasks as robots perform them, by searching the Internet, interpreting natural language text, images, and videos, object recognition as well as interaction. The project is led by Ashutosh Saxena at Stanford University.

RoboEarth is a project that has been described as a "World Wide Web for robots" – it is a network and database repository where robots can share information and learn from each other and a cloud for outsourcing heavy computation tasks. The project brings together researchers from five major universities in Germany, the Netherlands and Spain and is backed by the European Union.

Reinforcement Learning

Reinforcement learning (RL) is an area of machine learning inspired by behaviorist psychology, concerned with how software agents ought to take *actions* in an *environment* so as to maximize some notion of cumulative *reward*. The problem, due to its generality, is studied in many other disciplines, such as game theory, control theory, operations research, information theory, simulation-based optimization, multi-agent systems, swarm intelligence, statistics, and genetic algorithms. In the operations research and control literature, the field where reinforcement learning methods are studied is called *approximate dynamic programming*. The problem has been studied in the theory of optimal control, though most studies are concerned with the existence of optimal solutions and their characterization, and not with the learning or approximation aspects. In economics and game theory, reinforcement learning may be used to explain how equilibrium may arise under bounded rationality.

In machine learning, the environment is typically formulated as a Markov decision process (MDP), as many reinforcement learning algorithms for this context utilize dynamic programming techniques. The main difference between the classical techniques and reinforcement learning algorithms is that the latter do not need knowledge about the MDP and they target large MDPs where exact methods become infeasible.

Reinforcement learning differs from standard supervised learning in that correct input/output pairs are never presented, nor sub-optimal actions explicitly corrected. Further, there is a focus on on-line performance, which involves finding a balance between exploration (of uncharted territory) and exploitation (of current knowledge). The exploration vs. exploitation trade-off in reinforcement learning has been most thoroughly studied through the multi-armed bandit problem and in finite MDPs.

Introduction

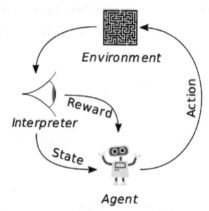

The typical framing of a Reinforcement Learning (RL) scenario: an agent takes actions in an environment which is interpreted into a reward and a representation of the state which is fed back into the agent.

The basic reinforcement learning model consists of:

1. a set of environment and agent states S;

2. a set of actions A of the agent;

3. policies of transitioning from states to actions;

4. rules that determine the *scalar immediate reward* of a transition; and

5. rules that describe what the agent observes.

The rules are often stochastic. The observation typically involves the scalar immediate reward associated with the last transition. In many works, the agent is also assumed to observe the current environmental state, in which case we talk about *full observability*, whereas in the opposing case we talk about *partial observability*. Sometimes the set of actions available to the agent is restricted (e.g., you cannot spend more money than what you possess).

A reinforcement learning agent interacts with its environment in discrete time steps. At each time t, the agent receives an observation o_t, which typically includes the reward r_t. It then chooses an action a_t from the set of actions available, which is subsequently sent to the environment. The environment moves to a new state s_{t+1} and the reward r_{t+1} associated with the *transition* (s_t, a_t, s_{t+1}) is determined. The goal of a reinforcement learning agent is to collect as much reward as possible. The agent can choose any action as a function of the history and it can even randomize its action selection.

When the agent's performance is compared to that of an agent which acts optimally from the beginning, the difference in performance gives rise to the notion of *regret*. Note that in order to act near optimally, the agent must reason about the long term consequences of its actions (i.e., maximize the future income), although the immediate reward associated with this might be negative.

Thus, reinforcement learning is particularly well-suited to problems which include a long-term versus short-term reward trade-off. It has been applied successfully to various problems, including robot control, elevator scheduling, telecommunications, backgammon, checkers (Sutton & Barto 1998, Chapter 11) and go (AlphaGo).

Two components make reinforcement learning powerful: The use of samples to optimize performance and the use of function approximation to deal with large environments. Thanks to these two key components, reinforcement learning can be used in large environments in any of the following situations:

- A model of the environment is known, but an analytic solution is not available;

- Only a simulation model of the environment is given (the subject of simulation-based optimization);

- The only way to collect information about the environment is by interacting with it.

The first two of these problems could be considered planning problems (since some form of the model is available), while the last one could be considered as a genuine learning problem. However, under a reinforcement learning methodology both planning problems would be converted to machine learning problems.

Exploration

The reinforcement learning problem as described requires clever exploration mechanisms. Randomly selecting actions, without reference to an estimated probability distribution, is known to give rise to very poor performance. The case of (small) finite Markov decision processes is relatively well understood by now. However, due to the lack of algorithms that would provably scale well with the number of states (or scale to problems with infinite state spaces), in practice people resort to simple exploration methods. One such method is ϵ-greedy, when the agent chooses the action that it believes has the best long-term effect with probability $1-\epsilon$, and it chooses an action uniformly at random, otherwise. Here, $0 < \epsilon < 1$ is a tuning parameter, which is sometimes changed, either according to a fixed schedule (making the agent explore less as time goes by), or adaptively based on some heuristics.

Algorithms for Control Learning

Even if the issue of exploration is disregarded and even if the state was observable (which we assume from now on), the problem remains to find out which actions are good based on past experience.

Criterion of Optimality

For simplicity, assume for a moment that the problem studied is *episodic*, an episode ending when some *terminal state* is reached. Assume further that no matter what course of actions the agent takes, termination is inevitable. Under some mild regularity conditions the expectation of the total reward is then well-defined, for *any* policy and any initial distribution over the states. Here, a policy means a map from states to probability distributions over actions:

$$\pi : S \rightarrow P(A = a \mid S)_.$$

Given a fixed initial distribution μ, we can thus assign the expected return ρ^{π} to policy π:

$$\rho^{\pi} = E[R \mid \pi],$$

where the random variable R denotes the *return* and is defined by

$$R = \sum_{t=0}^{N-1} r_{t+1},$$

where r_{t+1} is the reward received after the t-th transition, the initial state is sampled at random from μ and actions are selected by policy π. Here, N denotes the (random) time when a terminal state is reached, i.e., the time when the episode terminates.

In the case of non-episodic problems the return is often *discounted*,

$$R = \sum_{t=0}^{\infty} \gamma^{t} r_{t+1},$$

giving rise to the total expected discounted reward criterion. Here $0 \le \gamma \le 1$ is the so-called *discount-factor*. Since the undiscounted return is a special case of the discounted return, from now on we will assume discounting. Although this looks innocent enough, discounting is in fact problematic if one cares about online performance. This is because discounting makes the initial time steps more important. Since a learning agent is likely to make mistakes during the first few steps after its "life" starts, no uninformed learning algorithm can achieve near-optimal performance under discounting even if the class of environments is restricted to that of finite MDPs. (This does not mean though that, given enough time, a learning agent cannot figure how to act near-optimally, if time was restarted.)

The problem then is to specify an algorithm that can be used to find a policy with maximum expected return. From the theory of MDPs it is known that, without loss of generality, the search can be restricted to the set of the so-called *stationary* policies. A policy is called stationary if the action-distribution returned by it depends only on the last state visited (which is part of the observation history of the agent, by our simplifying assumption). In fact, the search can be further restricted to *deterministic* stationary policies. A deterministic stationary policy is one which deterministically selects actions based on the current state. Since any such policy can be identified with a mapping from the set of states to the set of actions, these policies can be identified with such mappings with no loss of generality.

Brute Force

The brute force approach entails the following two steps:

1. For each possible policy, sample returns while following it

2. Choose the policy with the largest expected return

One problem with this is that the number of policies can be extremely large, or even infinite. Another is that variance of the returns might be large, in which case a large number of samples will be required to accurately estimate the return of each policy.

These problems can be ameliorated if we assume some structure and perhaps allow samples generated from one policy to influence the estimates made for another. The two main approaches for achieving this are value function estimation and direct policy search.

Value Function Approaches

Value function approaches attempt to find a policy that maximizes the return by maintaining a set of estimates of expected returns for some policy (usually either the "current" (on-policy) or the optimal (off-policy) one).

These methods rely on the theory of MDPs, where optimality is defined in a sense which is stronger than the above one: A policy is called optimal if it achieves the best expected return from *any* initial state (i.e., initial distributions play no role in this definition). Again, one can always find an optimal policy amongst stationary policies.

To define optimality in a formal manner, define the value of a policy π by

$$V^{\pi}(s) = E[R \mid s, \pi],$$

where R stands for the random return associated with following π from the initial state s. Define $V^{*}(s)$ as the maximum possible value of $V^{*}(s)$, where π is allowed to change:

$$V^{*}(s) = \max_{\pi} V^{\pi}(s).$$

A policy which achieves these *optimal values* in *each* state is called *optimal*. Clearly, a policy that is optimal in this strong sense is also optimal in the sense that it maximizes the expected return ρ^{π}, since $\rho^{\pi} = E[V^{\pi}(S)]$, where S is a state randomly sampled from the distribution μ.

Although state-values suffice to define optimality, it will prove to be useful to define action-values. Given a state s, an action a and a policy π, the action-value of the pair (s, a) under π is defined by

$$Q^{\pi}(s, a) = E[R \mid s, a, \pi]$$

where, now, R stands for the random return associated with first taking action a in state s and following π, thereafter.

It is well-known from the theory of MDPs that if someone gives us Q for an optimal policy, we can always choose optimal actions (and thus act optimally) by simply choosing the action with the highest value at each state. The *action-value function* of such an optimal policy is called the *optimal action-value function* and is denoted by Q^{*}. In summary, the knowledge of the optimal action-value function *alone* suffices to know how to act optimally.

Assuming full knowledge of the MDP, there are two basic approaches to compute the optimal action-value function, value iteration and policy iteration. Both algorithms compute a sequence of functions Q_k ($k = 0, 1, 2, \ldots$) which converge to Q^*. Computing these functions involves computing expectations over the whole state-space, which is impractical for all but the smallest (finite) MDPs, never mind the case when the MDP is unknown. In reinforcement learning methods the expectations are approximated by averaging over samples and one uses function approximation techniques to cope with the need to represent value functions over large state-action spaces.

Monte Carlo Methods

The simplest Monte Carlo methods can be used in an algorithm that mimics policy iteration. Policy iteration consists of two steps: *policy evaluation* and *policy improvement*.

The Monte Carlo methods are used in the policy evaluation step. In this step, given a stationary, deterministic policy π, the goal is to compute the function values $Q^\pi(s, a)$ (or a good approximation to them) for all state-action pairs (s, a). Assume (for simplicity) that the MDP is finite and in fact a table representing the action-values fits into the memory. Further, assume that the problem is episodic and after each episode a new one starts from some random initial state. Then, the estimate of the value of a given state-action pair (s, a) can be computed by simply averaging the sampled returns which originated from (s, a) over time. Given enough time, this procedure can thus construct a precise estimate Q of the action-value function Q^π. This finishes the description of the policy evaluation step.

In the policy improvement step, as it is done in the standard policy iteration algorithm, the next policy is obtained by computing a *greedy* policy with respect to Q: Given a state s, this new policy returns an action that maximizes $Q(s, \cdot)$. In practice one often avoids computing and storing the new policy, but uses lazy evaluation to defer the computation of the maximizing actions to when they are actually needed.

A few problems with this procedure are as follows:

- The procedure may waste too much time on evaluating a suboptimal policy;

- It uses samples inefficiently in that a long trajectory is used to improve the estimate only of the *single* state-action pair that started the trajectory;

- When the returns along the trajectories have *high variance*, convergence will be slow;

- It works in *episodic problems only*;

- It works in *small, finite MDPs only*.

Temporal Difference Methods

The first issue is easily corrected by allowing the procedure to change the policy (at all, or at some states) before the values settle. However good this sounds, this may be problematic as this might prevent convergence. Still, most current algorithms implement this idea, giving rise to the class of *generalized policy iteration* algorithm. We note in passing that many actor critic methods belong to this category.

The second issue can be corrected within the algorithm by allowing trajectories to contribute to any state-action pair in them. This may also help to some extent with the third problem, although a better solution when returns have high variance is to Sutton's temporal difference (TD) methods which are based on the recursive Bellman equation. Note that the computation in TD methods can be incremental (when after each transition the memory is changed and the transition is thrown away), or batch (when the transitions are collected and then the estimates are computed once based on a large number of transitions). Batch methods, a prime example of which is the least-squares temporal difference method due to Bradtke & Barto 1996, may use the information in the samples better, whereas incremental methods are the only choice when batch methods become infeasible due to their high computational or memory complexity. In addition, there exist methods that try to unify the advantages of the two approaches. Methods based on temporal differences also overcome the second but last issue.

In *linear function approximation* one starts with a mapping ϕ that assigns a finite-dimensional vector to each state-action pair. Then, the action values of a state-action pair (s, a) are obtained by linearly combining the components of $\phi(s, a)$ with some *weights* θ:

$$Q(s, a) = \sum_{i=1}^{d} \theta_i \phi_i(s, a).$$

The algorithms then adjust the weights, instead of adjusting the values associated with the individual state-action pairs. However, linear function approximation is not the only choice. More recently, methods based on ideas from nonparametric statistics (which can be seen to construct their own features) have been explored.

So far, the discussion was restricted to how policy iteration can be used as a basis of the designing reinforcement learning algorithms. Equally importantly, value iteration can also be used as a starting point, giving rise to the Q-Learning algorithm (Watkins 1989) and its many variants.

The problem with methods that use action-values is that they may need highly precise estimates of the competing action values, which can be hard to obtain when the returns are noisy. Though this problem is mitigated to some extent by temporal difference methods and if one uses the so-called compatible function approximation method, more work remains to be done to increase generality and efficiency. Another problem specific to temporal difference methods comes from their reliance on the recursive Bellman equation. Most temporal difference methods have a so-called λ parameter ($0 \leq \lambda \leq 1$) that allows one to continuously interpolate between Monte-Carlo methods (which do not rely on the Bellman equations) and the basic temporal difference methods (which rely entirely on the Bellman equations), which can thus be effective in palliating this issue.

Direct Policy Search

An alternative method to find a good policy is to search directly in (some subset of) the policy space, in which case the problem becomes an instance of stochastic optimization. The two approaches available are gradient-based and gradient-free methods.

Gradient-based methods (giving rise to the so-called *policy gradient methods*) start with a mapping from a finite-dimensional (parameter) space to the space of policies: given the parameter vector θ, let π_θ denote the policy associated to θ. Define the performance function by

$$\rho(\theta) = \rho^{\pi_\theta}$$

Under mild conditions this function will be differentiable as a function of the parameter vector θ. If the gradient of ρ was known, one could use gradient ascent. Since an analytic expression for the gradient is not available, one must rely on a noisy estimate. Such an estimate can be constructed in many ways, giving rise to algorithms like Williams' REINFORCE method (which is also known as the likelihood ratio method in the simulation-based optimization literature). Policy gradient methods have received a lot of attention in the last couple of years (e.g., Peters, Vijayakumar & Schaal 2003), but they remain an active field. An overview of policy search methods in the context of robotics has been given by Deisenroth, Neumann and Peters. The issue with many of these methods is that they may get stuck in local optima (as they are based on local search).

A large class of methods avoids relying on gradient information. These include simulated annealing, cross-entropy search or methods of evolutionary computation. Many gradient-free methods can achieve (in theory and in the limit) a global optimum. In a number of cases they have indeed demonstrated remarkable performance.

The issue with policy search methods is that they may converge slowly if the information based on which they act is noisy. For example, this happens when in episodic problems the trajectories are long and the variance of the returns is large. As argued beforehand, value-function based methods that rely on temporal differences might help in this case. In recent years, several *actor–critic methods* have been proposed following this idea and were demonstrated to perform well in various problems.

Theory

The theory for small, finite MDPs is quite mature. Both the asymptotic and finite-sample behavior of most algorithms is well-understood. As mentioned beforehand, algorithms with provably good online performance (addressing the exploration issue) are known.

The theory of large MDPs needs more work. Efficient exploration is largely untouched (except for the case of bandit problems). Although finite-time performance bounds appeared for many algorithms in the recent years, these bounds are expected to be rather loose and thus more work is needed to better understand the relative advantages, as well as the limitations of these algorithms.

For incremental algorithm asymptotic convergence issues have been settled. Recently, new incremental, temporal-difference-based algorithms have appeared which converge under a much wider set of conditions than was previously possible (for example, when used with arbitrary, smooth function approximation).

Current Research

Current research topics include: adaptive methods which work with fewer (or no) parameters under a large number of conditions, addressing the exploration problem in large MDPs, large-scale empirical evaluations, learning and acting under partial information (e.g., using Predictive State Representation), modular and hierarchical reinforcement learning, improving existing value-function and policy search methods, algorithms that work well with large (or continuous) action spaces, transfer learning, lifelong learning, efficient sample-based planning (e.g., based on Monte-Carlo tree search). Multiagent or Distributed Reinforcement Learning is also a topic of interest in current research. There is also a growing interest in real life applications of reinforcement learning.

Reinforcement learning algorithms such as TD learning are also being investigated as a model for Dopamine-based learning in the brain. In this model, the dopaminergic projections from the substantia nigra to the basal ganglia function as the prediction error. Reinforcement learning has also been used as a part of the model for human skill learning, especially in relation to the interaction between implicit and explicit learning in skill acquisition (the first publication on this application was in 1995-1996, and there have been many follow-up studies).

There are multiple applications of reinforcement learning to generate models and train them to play video games, such as Atari games. In these models, reinforcement learning finds the actions with the best reward at each play. This method is a widely used method in combination with deep neural networks to teach computers to play Atari video games.

Inverse Reinforcement Learning

In inverse reinforcement learning (IRL), no reward function is given. Instead, one tries to extract the reward function given an observed behavior from an expert. The idea is to mimic the observed behavior which is often optimal or close to optimal.

In apprenticeship learning, one assumes that an expert demonstrating the ideal behavior, and tries to recover the policy directly using the observations from the expert.

Computational Learning Theory

In computer science, computational learning theory (or just learning theory) is a subfield of Artificial Intelligence devoted to studying the design and analysis of machine learning algorithms.

Overview

Theoretical results in machine learning mainly deal with a type of inductive learning called supervised learning. In supervised learning, an algorithm is given samples that are labeled in some useful way. For example, the samples might be descriptions of mushrooms, and the labels could be whether or not the mushrooms are edible. The algorithm takes these previously labeled samples and uses them to induce a classifier. This classifier is a function that assigns labels to samples including the samples that have never been previously seen by the algorithm. The goal of the supervised learning algorithm is to optimize some measure of performance such as minimizing the number of mistakes made on new samples.

In addition to performance bounds, computational learning theory studies the time complexity and feasibility of learning. In computational learning theory, a computation is considered feasible if it can be done in polynomial time. There are two kinds of time complexity results:

- Positive results – Showing that a certain class of functions is learnable in polynomial time.

- Negative results – Showing that certain classes cannot be learned in polynomial time.

Negative results often rely on commonly believed, but yet unproven assumptions, such as:

- Computational complexity – P ≠ NP (the P versus NP problem);
- Cryptographic – One-way functions exist.

There are several different approaches to computational learning theory. These differences are based on making assumptions about the inference principles used to generalize from limited data. This includes different definitions of probability and different assumptions on the generation of samples. The different approaches include:

- Exact learning, proposed by Dana Angluin;
- Probably approximately correct learning (PAC learning), proposed by Leslie Valiant;
- VC theory, proposed by Vladimir Vapnik and Alexey Chervonenkis;
- Bayesian inference;
- Algorithmic learning theory, from the work of E. Mark Gold;
- Online machine learning, from the work of Nick Littlestone.

Computational learning theory has led to several practical algorithms. For example, PAC theory inspired boosting, VC theory led to support vector machines, and Bayesian inference led to belief networks (by Judea Pearl).

References

- B. Biggio, G. Fumera, and F. Roli. "Multiple classifier systems for robust classifier design in adversarial environments". International Journal of Machine Learning and Cybernetics, 1(1):27–41, 2010
- Wiebe, Nathan; Braun, Daniel; Lloyd, Seth (2012). "Quantum Algorithm for Data Fitting". Physical Review Letters. 109 (5): 050505. Bibcode:2012PhRvL.109e0505W. PMID 23006156. doi:10.1103/PhysRevLett.109.050505
- M. Kloft and P. Laskov. "Security analysis of online centroid anomaly detection". Journal of Machine Learning Research, 13:3647–3690, 2012
- Vorobeychik, Yevgeniy; Li, Bo (2014-01-01). Proceedings of the 2014 International Conference on Autonomous Agents and Multi-agent Systems. AAMAS '14. Richland, SC: International Foundation for Autonomous Agents and Multiagent Systems: 485–492. ISBN 9781450327381
- Schuld, Maria; Sinayskiy, Ilya; Petruccione, Francesco (2016). "Prediction by linear regression on a quantum computer". Physical Review A. 94 (2): 022342. Bibcode:2016PhRvA..94b2342S. doi:10.1103/PhysRevA.94.022342
- Schaffer, Amanda. "10 Breakthrough Technologies 2016: Robots That Teach Each Other". MIT Technology Review. Retrieved 4 January 2017
- Yoo, Seokwon; Bang, Jeongho; Lee, Changhyoup; Lee, Jinhyoung (2013). "A quantum speedup in machine learning: Finding a N-bit Boolean function for a classification". New Journal of Physics. 16 (10): 103014. Bibcode:2014NJPh...16j3014Y. arXiv:1303.6055. doi:10.1088/1367-2630/16/10/103014
- Aïmeur, Esma; Brassard, Gilles; Gambs, Sébastien (2013-02-01). "Quantum speed-up for unsupervised learning". Machine Learning. 90 (2): 261–287. ISSN 0885-6125. doi:10.1007/s10994-012-5316-5
- Wittek, Peter (2014). Quantum Machine Learning: What Quantum Computing Means to Data Mining. Academic Press. ISBN 978-0-12-800953-6

- "New Worldwide Network Lets Robots Ask Each Other Questions When They Get Confused". Popular Science. Retrieved 4 January 2017

- van Nieuwenburg, Evert; Liu, Ye-Hua; Huber, Sebastian (2017). "Learning phase transitions by confusion". Nature Physics. doi:10.1038/nphys4037

- Dunjko, V.; Friis, N.; Briegel, H. J. (2015-01-01). "Quantum-enhanced deliberation of learning agents using trapped ions". New Journal of Physics. 17 (2): 023006. ISSN 1367-2630. doi:10.1088/1367-2630/17/2/023006

- Servedio, Rocco A.; Gortler, Steven J. (2004). "Equivalences and Separations Between Quantum and Classical Learnability". SIAM Journal on Computing. 33 (5): 1067–1092. doi:10.1137/S0097539704412910

An Integrated Study of Artificial Neural Network

Artificial neural networks are computing systems that are motivated by biological neural networks. The system is based on units called artificial neurons. Information is sent from one neuron to another and the receiving neuron processes it. This chapter will provide an integrated understanding of artificial neural networks.

Artificial Neural Network

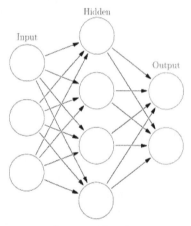

An artificial neural network is an interconnected group of nodes, akin to the vast network of neurons in a brain. Here, each circular node represents an artificial neuron and an arrow represents a connection from the output of one neuron to the input of another.

Artificial neural networks (ANNs) or connectionist systems are a computational model used in machine learning, computer science and other research disciplines.

Such systems can be trained from examples, rather than explicitly specified. They have found most use in applications difficult to express in a traditional computer algorithm using ordinary rule-based programming. Neural networks have been used to solve a wide variety of tasks, including computer vision, speech recognition, machine translation and medical diagnosis.

ANNs take their name from biological neural networks found in animal brains. Thusly, an ANN is based on a collection of connected units called artificial neurons, (analogous to axons in a biological brain). Each connection (synapse) between (two) neurons can transmit an unidirectional signal with an activating strength that varies with the strength of the connection. If the combined incoming signals (from potentially many transmitting neurons) are strong enough, the receiving (postsynaptic) neuron activates and propagates a signal to downstream neurons connected to it.

Typically, neurons are connected in layers, and signals travel from the first (input), to the last (output) layer. As of 2017 neural networks typically have a few thousand to a few million neural units and millions of connections; their computing power is similar to a worm brain, several orders of magnitude simpler than a human brain. The signals and state of artificial neurons are represented by real numbers, typically between 0 and 1. A threshold or limiting function may govern each connection and neuron, such that the signal must exceed the limit before propagating. Back propagation is the use of forward stimulation to modify connection weights. Training typically requires several thousand cycles of interaction.

The original goal of the neural network approach was to solve problems in the same way that a human brain would, although some varieties have become more abstract. ANNs potential value is that they can infer a function from sufficient observations. This is particularly useful where the complexity of the data or task makes the design of such a function by hand impracticable.

Brain research has repeatedly stimulated new ANN approaches. One new approach is the use of connections to connect processing layers rather than adjacent neurons. Other research explores the use of multiple signal types, or finer control than boolean (on/off) variables. Newer types of network allow connections to interact in more chaotic and complex ways. Dynamic neural networks can dynamically form new connections and even new neural units while disabling others.

Historically, the use of neural network models marked a directional shift in the late 1980s from high-level (symbolic) artificial intelligence, characterized by expert systems with knowledge embodied in *if-then* rules, to low-level (sub-symbolic) machine learning, characterized by knowledge embodied in the parameters of a cognitive model.

History

Warren McCulloch and Walter Pitts (1943) created a computational model for neural networks based on mathematics and algorithms called threshold logic. This model paved the way for neural network research to split into two distinct approaches. One approach focused on biological processes in the brain while the other focused on the application of neural networks to artificial intelligence. This work led work on nerve networks and their link to finite automata.

Hebbian Learning

In the late 1940s Hebb created a learning hypothesis based on the mechanism of neural plasticity that is now known as Hebbian learning. Hebbian learning is considered to be an unsupervised learning rule. Its later variants were models for long term potentiation. Researchers started applying these ideas to computational models in 1948 with Turing's B-type machines.

Farley and Clark (1954) first used computational machines, then called "calculators", to simulate a Hebbian network. Other neural network computational machines were created by Rochester, Holland, Habit and Duda (1956).

Rosenblatt (1958) created the perceptron, an algorithm for pattern recognition based on a two-layer network using simple addition and subtraction. With mathematical notation, Rosenblatt also described circuitry not in the basic perceptron, such as the exclusive-or circuit, which could not be processed by neural networks until after Werbos created the back propagation algorithm (1975).

Neural network research stagnated after the publication of machine learning research by Minsky and Papert (1969), who discovered two key issues with the computational machines that processed neural networks. The first was that basic perceptrons were incapable of processing the exclusive-or circuit. The second was that computers didn't have enough processing power to effectively handle the work required by large neural networks. Neural network research slowed until computers achieved far greater processing power.

Backpropagation and Resurgence

A key advance was the backpropagation algorithm that effectively solved the exclusive-or problem, and more generally accelerated the training of multi-layer networks (Werbos 1975).

In the mid-1980s, parallel distributed processing became popular under the name connectionism. The textbook by Rumelhart and McClelland (1986) described the use of connectionism in computers to simulate neural processes.

Support vector machines and other, much simpler methods such as linear classifiers gradually overtook neural networks in machine learning popularity. As earlier challenges in training deep neural networks were successfully addressed with methods such as unsupervised pre-training and available computing power increased through the use of GPUs and distributed computing, neural networks were deployed on a large scale, particularly in image and visual recognition problems. This became known as "deep learning", although deep learning is not strictly synonymous with deep neural networks.

Improvements Since 2006

Computational devices have been created in CMOS, for both biophysical simulation and neuromorphic computing. More recent efforts show promise for creating nanodevices for very large scale principal components analyses and convolution. If successful, these developments would create a new class of neural computing that depends on learning rather than programming and because it is fundamentally analog rather than digital even though the first implementations may use digital devices.

Between 2009 and 2012, recurrent neural networks and deep feedforward neural networks developed in the research group of Schmidhuber who won eight international competitions in pattern recognition and machine learning. For example, the bi-directional and multi-dimensional long short-term memory (LSTM) of Graves et al. won three competitions in connected handwriting recognition at the 2009 International Conference on Document Analysis and Recognition (ICDAR), without any prior knowledge about the three languages to be learned.

Fast GPU-based implementations of this approach by Ciresan and colleagues won several pattern recognition contests, including the IJCNN 2011 Traffic Sign Recognition Competition, the ISBI 2012 Segmentation of Neuronal Structures in Electron Microscopy Stacks challenge and others. Their neural networks were the first pattern recognizers to achieve human-competitive or even superhuman performance on important benchmarks such as traffic sign recognition (IJCNN 2012), or the MNIST handwritten digits problem.

Deep, highly nonlinear neural architectures similar to the 1980 neocognitron by Fukushima and the "standard architecture of vision", inspired by the simple and complex cells identified by Hubel and Wiesel in the primary visual cortex, were pre-trained by unsupervised methods by Hinton. A team from his lab won a 2012 contest sponsored by Merck to design software to help find molecules that might lead to new drugs.

Models

Neural network models are essentially simple mathematical models defining a function $f : X \rightarrow Y$ or a distribution over X or both X and Y. Sometimes models are intimately associated with a particular learning rule. A common use of the phrase "ANN model" is really the definition of a *class* of such functions (where members of the class are obtained by varying parameters, connection weights, or specifics of the architecture such as the number of neurons or their connectivity).

Network Function

The word *network* in the term 'artificial neural network' refers to the connections between the neurons. An example system has three layers. The first layer has input neurons which send data via synapses to the second layer of neurons, and then via more synapses to the third layer of output neurons. More complex systems have more layers of neurons. Some systems have more layers of input neurons and output neurons. An ANN is typically defined by three types of parameters:

- The connection pattern between the different layers of neurons

- The weights of the connections, which are updated in the learning process.

- The activation function that converts a neuron's weighted input to its output activation.

Mathematically, a neuron's network function $f(x)$ is defined as a composition of other functions $g_i(x)$, that can further be decomposed into other functions. This can be conveniently represented as a network structure, with arrows depicting the dependencies between functions. A widely used type of composition is the *nonlinear weighted sum*, where $f(x) = K\left(\sum_i w_i g_i(x)\right)$, where K (commonly referred to as the activation function) is some predefined function, such as the hyperbolic tangent or sigmoid function. The important characteristic of the activation function is that it provides a smooth transition as input values change, i.e. a small change in input produces a small change in output. The following refers to a collection of functions g_i as a vector $g = (g_1, g_2, \ldots, g_n)$.

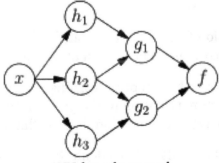

ANN dependency graph

This figure depicts such a decomposition of f, with dependencies between variables indicated by arrows. These can be interpreted in two ways.

The first view is the functional view: the input x is transformed into a 3-dimensional vector h, which is then transformed into a 2-dimensional vector g, which is finally transformed into f. This view is most commonly encountered in the context of optimization.

The second view is the probabilistic view: the random variable $F = f(G)$ depends upon the random variable $G = g(H)$, which depends upon $H = h(X)$, which depends upon the random variable X. This view is most commonly encountered in the context of graphical models.

The two views are largely equivalent. In either case, for this particular architecture, the components of individual layers are independent of each other (e.g., the components of g are independent of each other given their input h). This naturally enables a degree of parallelism in the implementation.

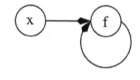

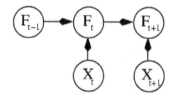

Two separate depictions of the recurrent ANN dependency graph.

Networks such as the previous one are commonly called feedforward, because their graph is a directed acyclic graph. Networks with cycles are commonly called recurrent. Such networks are commonly depicted in the manner shown at the top of the figure, where f is shown as being dependent upon itself. However, an implied temporal dependence is not shown.

Learning

The possibility of learning has attracted the most interest in neural networks. Given a specific *task* to solve, and a class of functions F, learning means using a set of observations to find $f^* \in F$ which solves the task in some optimal sense.

This entails defining a cost function $C : F \to \mathbb{R}$ such that, for the optimal solution f^*, $C(f^*) \le C(f)$ $\forall f \in F$ – i.e., no solution has a cost less than the cost of the optimal solution.

The cost function C is an important concept in learning, as it is a measure of how far away a particular solution is from an optimal solution to the problem to be solved. Learning algorithms search through the solution space to find a function that has the smallest possible cost.

For applications where the solution is data dependent, the cost must necessarily be a function of the observations, otherwise the model would not relate to the data. It is frequently defined as a statistic to which only approximations can be made. As a simple example, consider the problem of finding the model f, which minimizes $C = E\left[(f(x) - y)^2 \right]$, for data pairs (x, y) drawn from some distribution $\mathcal{D}$. In practical situations we would only have N samples from $\mathcal{D}$ and thus, for the above example, we would only minimize $\hat{C} = \frac{1}{N}\sum_{i=1}^{N}(f(x_i) - y_i)^2$. Thus, the cost is minimized over a sample of the data rather than the entire distribution.

When $N \to \infty$ some form of online machine learning must be used, where the cost is reduced as each new example is seen. While online machine learning is often used when $\mathcal{D}$ is fixed, it is most useful in the case where the distribution changes slowly over time. In neural network methods, some form of online machine learning is frequently used for finite datasets.

Choosing a Cost Function

While it is possible to define an ad hoc cost function, frequently a particular cost (function) is used, either because it has desirable properties (such as convexity) or because it arises naturally from a particular formulation of the problem (e.g., in a probabilistic formulation the posterior probability of the model can be used as an inverse cost). Ultimately, the cost function depends on the task.

Learning Paradigms

The three major learning paradigms each correspond to a particular learning task. These are supervised learning, unsupervised learning and reinforcement learning.

Supervised Learning

Supervised learning uses a set of example pairs $(x, y), x \in X, y \in Y$ and the aim is to find a function $f : X \to Y$ in the allowed class of functions that matches the examples. In other words, we wish to *infer* the mapping implied by the data; the cost function is related to the mismatch between our mapping and the data and it implicitly contains prior knowledge about the problem domain.

A commonly used cost is the mean-squared error, which tries to minimize the average squared error between the network's output, $f(x)$, and the target value y over all the example pairs. Minimizing this cost using gradient descent for the class of neural networks called multilayer perceptrons (MLP), producs the backpropagation algorithm for training neural networks.

Tasks that fall within the paradigm of supervised learning are pattern recognition (also known as classification) and regression (also known as function approximation). The supervised learning paradigm is also applicable to sequential data (e.g., for speech and gesture recognition). This can be thought of as learning with a "teacher", in the form of a function that provides continuous feedback on the quality of solutions obtained thus far.

Unsupervised Learning

In unsupervised learning, some data x is given and the cost function to be minimized, that can be any function of the data x and the network's output, f.

The cost function is dependent on the task (the model domain) and any *a priori* assumptions (the implicit properties of the model, its parameters and the observed variables).

As a trivial example, consider the model $f(x) = a$ where a is a constant and the cost $C = E[(x - f(x))^2]$. Minimizing this cost produces a value of a that is equal to the mean of the data. The cost function can be much more complicated. Its form depends on the application: for example, in compression it could be related to the mutual information between x and $f(x)$, whereas in statistical modeling, it could be related to the posterior probability of the model given the data (note that in both of those examples those quantities would be maximized rather than minimized).

Tasks that fall within the paradigm of unsupervised learning are in general estimation problems; the applications include clustering, the estimation of statistical distributions, compression and filtering.

Reinforcement Learning

In reinforcement learning, data x are usually not given, but generated by an agent's interactions with the environment. At each point in time t, the agent performs an action y_t and the environment generates an observation x_t and an instantaneous cost c_t, according to some (usually unknown) dynamics. The aim is to discover a policy for selecting actions that minimizes some measure of a long-term cost, e.g., the expected cumulative cost. The environment's dynamics and the long-term cost for each policy are usually unknown, but can be estimated.

More formally the environment is modeled as a Markov decision process (MDP) with states $s_1,...,s_n \in S$ and actions $a_1,...,a_m \in A$ with the following probability distributions: the instantaneous cost distribution $P(c_t | s_t)$, the observation distribution $P(x_t | s_t)$ and the transition $P(s_{t+1} | s_t, a_t)$, while a policy is defined as the conditional distribution over actions given the observations. Taken together, the two then define a Markov chain (MC). The aim is to discover the policy (i.e., the MC) that minimizes the cost.

ANNs are frequently used in reinforcement learning as part of the overall algorithm. Dynamic programming was coupled with ANNs (giving neurodynamic programming) by Bertsekas and Tsitsiklis and applied to multi-dimensional nonlinear problems such as those involved in vehicle routing, natural resources management or medicine because of the ability of ANNs to mitigate losses of accuracy even when reducing the discretization grid density for numerically approximating the solution of the original control problems.

Tasks that fall within the paradigm of reinforcement learning are control problems, games and other sequential decision making tasks.

Learning Algorithms

Training a neural network model essentially means selecting one model from the set of allowed models (or, in a Bayesian framework, determining a distribution over the set of allowed models) that minimizes the cost. Numerous algorithms are available for training neural network models; most of them can be viewed as a straightforward application of optimization theory and statistical estimation.

Most employ some form of gradient descent, using backpropagation to compute the actual gradients. This is done by simply taking the derivative of the cost function with respect to the network parameters and then changing those parameters in a gradient-related direction. Bckpropagation training algorithms fall into three categories:

- steepest descent (with variable learning rate and momentum, resilient backpropagation);

- quasi-Newton (Broyden-Fletcher-Goldfarb-Shanno, one step secant);

- Levenberg-Marquardt and conjugate gradient (Fletcher-Reeves update, Polak-Ribiére update, Powell-Beale restart, scaled conjugate gradient).

Evolutionary methods, gene expression programming, simulated annhealing, expectation-maximization, non-parametric methods and particle swarm optimization are other methods for training neural networks.

Use

Using ANNs requires an understanding of their characteristics.

- Choice of model: This depends on the data representation and the application. Overly complex models slow learning.

- Learning algorithm: Numerous trade-offs exist between learning algorithms. Almost any algorithm will work well with the correct hyperparameters for training on a particular data set. However, selecting and tuning an algorithm for training on unseen data requires significant experimentation.

- Robustness: If the model, cost function and learning algorithm are selected appropriately, the resulting ANN can become robust.

ANN capabilities fall within the following broad categories:

- Function approximation, or regression analysis, including time series prediction, fitness approximation and modeling.

- Classification, including pattern and sequence recognition, novelty detection and sequential decision making.

- Data processing, including filtering, clustering, blind source separation and compression.

- Robotics, including directing manipulators and prostheses.

- Control, including computer numerical control.

Applications

Application areas include system identification and control (vehicle control, trajectory prediction, process control, natural resources management), quantum chemistry, game-playing and decision making (backgammon, chess, poker), pattern recognition (radar systems, face identification, signal classification, object recognition and more), sequence recognition (gesture, speech, handwritten text recognition), medical diagnosis, finance (e.g. automated trading systems), data mining, visualization, machine translation, social network filtering and e-mail spam filtering.

ANNs have been used to diagnose cancers, including lung cancer, prostate cancer, colorectal cancer and to distinguish highly invasive cancer cell lines from less invasive lines using only cell shape information.

Neuroscience

Theoretical and computational neuroscience is concerned with the theoretical analysis and the computational modeling of biological neural systems. Since neural systems attempt to reflect cognitive processes and behavior, the field is closely related to cognitive and behavioral modeling.

To gain this understanding, neuroscientists strive to link observed biological processes (data), biologically plausible mechanisms for neural processing and learning (biological neural network models) and theory (statistical learning theory and information theory).

Types of Models

Many types of models are used, defined at different levels of abstraction and modeling different aspects of neural systems. They range from models of the short-term behavior of individual neurons, models of how the dynamics of neural circuitry arise from interactions between individual neurons and finally to models of how behavior can arise from abstract neural modules that represent complete subsystems. These include models of the long-term, and short-term plasticity, of neural systems and their relations to learning and memory from the individual neuron to the system level.

Networks with Memory

Integrating external memory components with artificial neural networks dates to early research in distributed representations and self-organizing maps. E.g. in sparse distributed memory the patterns encoded by neural networks are used as memory addresses for content-addressable memory, with "neurons" essentially serving as address encoders and decoders.

More recently deep learning was shown to be useful in semantic hashing where a deep graphical model of the word-count vectors is obtained from a large document set. Documents are mapped to memory addresses in such a way that semantically similar documents are located at nearby addresses. Documents similar to a query document can then be found by simply accessing other nearby addresses.

Memory networks are another extension to neural networks incorporating long-term memory. Long-term memory can be read and written to, with the goal of using it for prediction. These models have been applied in the context of question answering (QA) where the long-term memory effectively acts as a knowledge base, and the output is a textual response.

Neural turing machines (NTM) extend the capabilities of deep neural networks by coupling them to external memory resources, which they can interact with by attentional processes. The combined system is analogous to a Turing Machine but is differentiable end-to-end, allowing it to be efficiently trained with gradient descent. Preliminary results demonstrate that NTMs can infer simple algorithms such as copying, sorting and associative recall from input and output examples.

Differentiable neural computers (DNC) are an NTM extension. They out-performed Neural turing machines, long short-term memory systems and memory networks on sequence-processing tasks.

Theoretical Properties

Computational Power

The multilayer perceptron is a universal function approximator, as proven by the universal approximation theorem. However, the proof is not constructive regarding the number of neurons required, the network topology, the weights and the learning parameters.

A specific recurrent architecture with rational valued weights (as opposed to full precision real number-valued weights) has the full power of a universal Turing machine, using a finite number of neurons and standard linear connections. Further, the use of irrational values for weights results in a machine with super-Turing power.

Capacity

Models' "capacity" property roughly corresponds to their ability to model any given function. It is related to the amount of information that can be stored in the network and to the notion of complexity.

Convergence

Models may not consistently converge on a single solution, firstly because many local minima may exist, depending on the cost function and the model. Secondly, the optimization method used might not guarantee to converge when it begins far from any local minimum. Thirdly, for sufficiently large data or parameters, some methods become impractical.

Generalization and Statistics

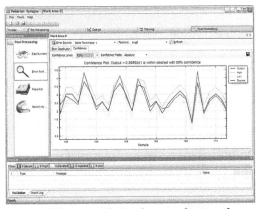

Confidence analysis of a neural network

Applications whose goal is to create a system that generalizes well to unseen examples, face the possibility of over-training. This arises in convoluted or over-specified systems when the capacity of the network significantly exceeds the needed free parameters. Two approaches address over-training. The first is to use cross-validation and similar techniques to check for the presence of over-training and optimally select hyperparameters to minimize the generalization error. The second is to use some form of *regularization*. This concept emerges in a probabilistic (Bayesian) framework, where regularization can be performed by selecting a larger prior probability over simpler models; but also in statistical learning theory, where the goal is to minimize over two quantities: the 'empirical risk' and the 'structural risk', which roughly corresponds to the error over the training set and the predicted error in unseen data due to overfitting.

Supervised neural networks that use a mean squared error (MSE) cost function can use formal statistical methods to determine the confidence of the trained model. The MSE on a validation set can be used as an estimate for variance. This value can then be used to calculate the confidence interval of the output of the network, assuming a normal distribution. A confidence analysis made this way is statistically valid as long as the output probability distribution stays the same and the network is not modified.

By assigning a softmax activation function, a generalization of the logistic function, on the output layer of the neural network (or a softmax component in a component-based neural network) for categorical target variables, the outputs can be interpreted as posterior probabilities. This is very useful in classification as it gives a certainty measure on classifications.

The softmax activation function is:

$$y_i = \frac{e^{x_i}}{\sum_{j=1}^{c} e^{x_j}}$$

Criticism

Training Issues

A common criticism of neural networks, particularly in robotics, is that they require too much training for real-world operation. Potential solutions include randomly shuffling training examples, by using a numerical optimization algorithm that does not take too large steps when changing the network connections following an example and by grouping examples in so-called mini-batches.

Theoretical Issues

No neural network has solved such computationally difficult problems such as the n-Queens problem, the travelling salesman problem, or the problem of factoring large integers.

A fundamental objection is that they do not reflect how real neurons function. Back propagation is a critical part of most artificial neural networks, although no such mechanism exists in biological neural networks. How information is coded by real neurons is not known. Sensor neurons fire action potentials more frequently with sensor activation and muscle cells pull more strongly when their associated motor neurons receive action potentials more frequently. Other than the case of relaying information from a sensor neuron to a motor neuron, almost nothing of the principles of how information is handled by biological neural networks is known.

The motivation behind ANNs is not necessarily to strictly replicate neural function, but to use biological neural networks as an inspiration. A central claim of ANNs is therefore that it embodies some new and powerful general principle for processing information. Unfortunately, these general principles are ill-defined. It is often claimed that they are emergent from the network itself. This allows simple statistical association (the basic function of artificial neural networks) to be described as learning or recognition. As a result, artificial neural networks have a "something-for-nothing quality, one that imparts a peculiar aura of laziness and a distinct lack of curiosity about just how good these computing systems are. No human hand (or mind) intervenes; solutions are found as if by magic; and no one, it seems, has learned anything".

Hardware Issues

Large and effective software neural networks considerable computing resources. While the brain has hardware tailored to the task of processing signals through a graph of neurons, simulating even a simplified neuron on von Neumann architecture may compel a neural network designer to fill many millions of database rows for its connections – which can consume vast amounts of memory and storage. Furthermore, the designer often needs to transmit signals through many of these connections and their associated neurons – which must often be matched with enormous CPU processing power and time.

Schmidhuber notes that the resurgence of neural networks in the twenty-first century is largely attributable to advances in hardware: from 1991 to 2015, computing power, especially as delivered by GPGPUs (on GPUs), has increased around a million-fold, making the standard backpropagation algorithm feasible for training networks that are several layers deeper than before. The use of parallel GPUs can reduce training times from months to days.

Neuromorphic engineering addresses the hardware difficulty directly, by constructing non-von-Neumann chips to directly implement neural networks in circuitry. Another chip optimized for neural network processing is called a Tensor Processing Unit, or TPU.

Practical Counterexamples to Criticisms

Arguments against Dewdney's position are that neural networks have been successfully used to solve many complex and diverse tasks, ranging from autonomously flying aircraft to detecting credit card fraud to mastering the game of Go.

Technology writer Roger Bridgman commented:

Neural networks, for instance, are in the dock not only because they have been hyped to high heaven, (what hasn't?) but also because you could create a successful net without understanding how it worked: the bunch of numbers that captures its behaviour would in all probability be "an opaque, unreadable table...valueless as a scientific resource".

In spite of his emphatic declaration that science is not technology, Dewdney seems here to pillory neural nets as bad science when most of those devising them are just trying to be good engineers. An unreadable table that a useful machine could read would still be well worth having.

Although it is true that analyzing what has been learned by an artificial neural network is difficult, it is much easier to do so than to analyze what has been learned by a biological neural network. Furthermore, researchers involved in exploring learning algorithms for neural networks are gradually uncovering general principles that allow a learning machine to be successful. For example, local vs non-local learning and shallow vs deep architecture.

Hybrid Approaches

Advocates of hybrid models (combining neural networks and symbolic approaches), claim that such a mixture can better capture the mechanisms of the human mind.

Types

Artificial neural network types vary from those with only one or two layers of single direction logic, to complicated multi–input many directional feedback loops and layers. On the whole, these systems use algorithms in their programming to control and organize their functions. Most systems use "weights" to change the parameters of the throughput and the varying connections to the neurons. Artificial neural networks can be autonomous and learn by input from outside "teachers" or teach themselves from written-in rules. Neural Cube style neural networks provide a dynamic space in which networks dynamically recombine information and links across self adapting nodes utilizing Neural Darwinism, which allows for more biologically modeled systems.

Types of Artificial Neural Networks

Artificial neural networks are computational models inspired by biological neural networks, and are used to approximate functions that are generally unknown. Particularly, they are inspired by the behaviour of neurons and the electrical signals they convey between input (such as from the eyes or nerve endings in the hand), processing, and output from the brain (such as reacting to light, touch, or heat). The way neurons semantically communicate is an area of ongoing research. Most artificial neural networks bear only some resemblance to their more complex biological counterparts, but are very effective at their intended tasks (e.g. classification or segmentation).

Some ANNs are adaptive systems and are used for example to model populations and environments, which constantly change.

Neural networks can be hardware- (neurons are represented by physical components) or software-based (computer models), and can use a variety of topologies and learning algorithms.

Feedforward Neural Network

The feedforward neural network was the first and arguably most simple type of artificial neural network devised. In this network the information moves in only one direction—forward: From the input nodes data goes through the hidden nodes (if any) and to the output nodes. There are no cycles or loops in the network. Feedforward networks can be constructed from different types of units, e.g. binary McCulloch-Pitts neurons, the simplest example being the perceptron. Continuous neurons, frequently with sigmoidal activation, are used in the context of backpropagation of error.

General Description of RBF Networks

Radial basis functions are powerful techniques for interpolation in multidimensional space. A RBF is a function which has built into it a distance criterion with respect to a center. Radial basis functions have been applied in the area of neural networks where they may be used as a replacement for the sigmoidal hidden layer transfer characteristic in multi-layer perceptrons. RBF networks have two layers of processing: In the first, input is mapped onto each RBF in the 'hidden' layer. The RBF chosen is usually a Gaussian. In regression problems the output layer is then a linear combination of hidden layer values representing mean predicted output. The interpretation of this output layer value is the same as a regression model in statistics. In classification problems the output layer is typically a sigmoid function of a linear combination of hidden layer values, representing a posterior probability. Performance in both cases is often improved by shrinkage techniques, known as ridge regression in classical statistics and known to correspond to a prior belief in small parameter values (and therefore smooth output functions) in a Bayesian framework.

RBF networks have the advantage of not suffering from local minima in the same way as multi-layer perceptrons. This is because the only parameters that are adjusted in the learning process are the linear mapping from hidden layer to output layer. Linearity ensures that the error surface is quadratic and therefore has a single easily found minimum. In regression problems this can be found in one matrix operation. In classification problems the fixed non-linearity introduced by the sigmoid output function is most efficiently dealt with using iteratively re-weighted least squares.

RBF networks have the disadvantage of requiring good coverage of the input space by radial basis functions. RBF centres are determined with reference to the distribution of the input data, but without reference to the prediction task. As a result, representational resources may be wasted on areas of the input space that are irrelevant to the learning task. A common solution is to associate each data point with its own centre, although this can make the linear system to be solved in the final layer rather large, and requires shrinkage techniques to avoid overfitting.

Associating each input datum with an RBF leads naturally to kernel methods such as support vector machines and Gaussian processes (the RBF is the kernel function). All three approaches use a non-linear kernel function to project the input data into a space where the learning problem can be solved using a linear model. Like Gaussian Processes, and unlike SVMs, RBF networks are typically trained in a Maximum Likelihood framework by maximizing the probability (minimizing the error) of the data under the model. SVMs take a different approach to avoiding overfitting by maximizing instead a margin. RBF networks are outperformed in most classification applications by SVMs. In regression applications they can be competitive when the dimensionality of the input space is relatively small.

How RBF Networks Work

Although the implementation is very different, RBF neural networks are conceptually similar to K-Nearest Neighbor (k-NN) models. The basic idea is that a predicted target value of an item is likely to be about the same as other items that have close values of the predictor variables. Consider this figure:

Assume that each case in the training set has two predictor variables, x and y. The cases are plotted using their x,y coordinates as shown in the figure. Also assume that the target variable has two categories, positive which is denoted by a square and negative which is denoted by a dash. Now, suppose we are trying to predict the value of a new case represented by the triangle with predictor values x=6, y=5.1. Should we predict the target as positive or negative?

Notice that the triangle is position almost exactly on top of a dash representing a negative value. But that dash is in a fairly unusual position compared to the other dashes which are clustered below the squares and left of center. So it could be that the underlying negative value is an odd case.

The nearest neighbor classification performed for this example depends on how many neighboring points are considered. If 1-NN is used and only the closest point is considered, then clearly the new point should be classified as negative since it is on top of a known negative point. On the other hand, if 9-NN classification is used and the closest 9 points are considered, then the effect of the surrounding 8 positive points may overbalance the close negative point.

An RBF network positions one or more RBF neurons in the space described by the predictor variables (x,y in this example). This space has as many dimensions as there are predictor variables. The Euclidean distance is computed from the point being evaluated (e.g., the triangle in this figure) to the center of each neuron, and a radial basis function (RBF) (also called a kernel function) is applied to the distance to compute the weight (influence) for each neuron. The radial basis function is so named because the radius distance is the argument to the function.

Weight = RBF(*distance*)

The further a neuron is from the point being evaluated, the less influence it has.

Radial Basis Function

Different types of radial basis functions could be used, but the most common is the Gaussian function:

If there is more than one predictor variable, then the RBF function has as many dimensions as there are variables. The following picture illustrates three neurons in a space with two predictor variables, X and Y. Z is the value coming out of the RBF functions:

The best predicted value for the new point is found by summing the output values of the RBF functions multiplied by weights computed for each neuron.

The radial basis function for a neuron has a center and a radius (also called a spread). The radius may be different for each neuron, and, in RBF networks generated by DTREG, the radius may be different in each dimension.

With larger spread, neurons at a distance from a point have a greater influence.

RBF Network Architecture

RBF networks have three layers:

1. Input layer: There is one neuron in the input layer for each predictor variable. In the case of categorical variables, N-1 neurons are used where N is the number of categories. The input neurons (or processing before the input layer) standardizes the range of the values by subtracting the median and dividing by the interquartile range. The input neurons then feed the values to each of the neurons in the hidden layer.

2. Hidden layer: This layer has a variable number of neurons (the optimal number is determined by the training process). Each neuron consists of a radial basis function centered on a point with as many dimensions as there are predictor variables. The spread (radius) of the RBF function may be different for each dimension. The centers and spreads are determined by the training process. When presented with the x vector of input values from the input layer, a hidden neuron computes the Euclidean distance of the test case from the neuron's center point and then applies the RBF kernel function to this distance using the spread values. The resulting value is passed to the summation layer.

3. Summation layer: The value coming out of a neuron in the hidden layer is multiplied by a weight associated with the neuron (W1, W2, ...,Wn in this figure) and passed to the summation which adds up the weighted values and presents this sum as the output of the network. Not shown in this figure is a bias value of 1.0 that is multiplied by a weight Wo and fed into the summation layer. For classification problems, there is one output (and a separate set of weights and summation unit) for each target category. The value output for a category is the probability that the case being evaluated has that category.

Training RBF Networks

The following parameters are determined by the training process:

- The number of neurons in the hidden layer

- The coordinates of the center of each hidden-layer RBF function

- The radius (spread) of each RBF function in each dimension

- The weights applied to the RBF function outputs as the≥°°y are passed to the summation layer

Various methods have been used to train RBF networks. One approach first uses K-means clustering to find cluster centers which are then used as the centers for the RBF functions. However, K-means clustering is a computationally intensive procedure, and it often does not generate the optimal number of centers. Another approach is to use a random subset of the training points as the centers.

DTREG uses a training algorithm developed by Sheng Chen, Xia Hong and Chris J. Harris. This algorithm uses an evolutionary approach to determine the optimal center points and spreads for each neuron. It also determines when to stop adding neurons to the network by monitoring the estimated leave-one-out (LOO) error and terminating when the LOO error begins to increase because of overfitting.

The computation of the optimal weights between the neurons in the hidden layer and the summation layer is done using ridge regression. An iterative procedure developed by Mark Orr (Orr, 1966) is used to compute the optimal regularization Lambda parameter that minimizes the generalized cross-validation (GCV) error.

Recurrent Neural Network

Contrary to feedforward networks, recurrent neural networks (RNNs) are models with bi-directional data flow. While a feedforward network propagates data linearly from input to output, RNNs also propagate data from later processing stages to earlier stages. RNNs can be used as general sequence processors.

Fully Recurrent Network

This is the basic architecture developed in the 1980s: a network of neuron-like units, each with a directed connection to every other unit. Each unit has a time-varying real-valued (more than just zero or one) activation (output). Each connection has a modifiable real-valued weight. Some of the nodes are called input nodes, some output nodes, the rest hidden nodes. Most architectures below are special cases.

For supervised learning in discrete time settings, training sequences of real-valued input vectors become sequences of activations of the input nodes, one input vector at a time. At any given time step, each non-input unit computes its current activation as a nonlinear function of the weighted sum of the activations of all units from which it receives connections. There may be teacher-given target activations for some of the output units at certain time steps. For example, if the input sequence is a speech signal corresponding to a spoken digit, the final target output at the end of the sequence may be a label classifying the digit. For each sequence, its error is the sum of the deviations of all activations computed by the network from the corresponding target signals. For a training set of numerous sequences, the total error is the sum of the errors of all individual sequences.

To minimize total error, gradient descent can be used to change each weight in proportion to its derivative with respect to the error, provided the non-linear activation functions are differentiable. Various methods for doing so were developed in the 1980s and early 1990s by Paul Werbos, Ronald J. Williams, Tony Robinson, Jürgen Schmidhuber, Barak Pearlmutter, and others. The standard method is called "backpropagation through time" or BPTT, a generalization of back-propagation for feedforward networks. A more computationally expensive online variant is called "Real-Time Recurrent Learning" or RTRL. Unlike BPTT this algorithm is *local in time but not local in space*. There also is an online hybrid between BPTT and RTRL with intermediate complexity, and there are variants for continuous time. A major problem with gradient descent for standard RNN architectures is that error gradients vanish exponentially quickly with the size of the time lag between important events, as first realized by Sepp Hochreiter in 1991. The Long short-term memory architecture overcomes these problems.

In reinforcement learning settings, there is no teacher providing target signals for the RNN, instead a fitness function or reward function or utility function is occasionally used to evaluate the performance of the RNN, which is influencing its input stream through output units connected to actuators affecting the environment. Variants of evolutionary computation are often used to optimize the weight matrix.

Hopfield Network

The Hopfield network (like similar attractor-based networks) is of historic interest although it is not a general RNN, as it is not designed to process sequences of patterns. Instead it requires stationary inputs. It is an RNN in which all connections are symmetric. Invented by John Hopfield in

1982 it guarantees that its dynamics will converge. If the connections are trained using Hebbian learning then the Hopfield network can perform as robust content-addressable memory, resistant to connection alteration.

Boltzmann Machine

The Boltzmann machine can be thought of as a noisy Hopfield network. Invented by Geoff Hinton and Terry Sejnowski in 1985, the Boltzmann machine is important because it is one of the first neural networks to demonstrate learning of latent variables (hidden units). Boltzmann machine learning was at first slow to simulate, but the contrastive divergence algorithm of Geoff Hinton (circa 2000) allows models such as Boltzmann machines and Products of Experts to be trained much faster.

Self-organizing Map

The self-organizing map (SOM) invented by Teuvo Kohonen performs a form of unsupervised learning. A set of artificial neurons learn to map points in an input space to coordinates in an output space. The input space can have different dimensions and topology from the output space, and the SOM will attempt to preserve these.

Learning Vector Quantization

Learning vector quantization (LVQ) can also be interpreted as a neural network architecture. It was originally suggested by Teuvo Kohonen. In LVQ, prototypical representatives of the classes parameterize, together with an appropriate distance measure, a distance-based classification scheme.

Simple Recurrent Networks

A simple modification of the basic feedforward architecture above was employed by Jeff Elman and Michael I. Jordan. A three-layer network is used, with the addition of a set of "context units" in the input layer. There are connections from the hidden layer (Elman) or from the output layer (Jordan) to these context units fixed with a weight of one. At each time step, the input is propagated in a standard feedforward fashion, and then a simple backpropagation-like learning rule is applied (this rule is not performing proper gradient descent, however). The fixed back connections result in the context units always maintaining a copy of the previous values of the hidden units (since they propagate over the connections before the learning rule is applied).

Echo State Network

The echo state network (ESN) is a recurrent neural network with a sparsely connected random hidden layer. The weights of output neurons are the only part of the network that can change and be trained. ESN are good at reproducing certain time series. A variant for spiking neurons is known as Liquid state machines.

Long Short-term Memory Network

The long short-term memory (LSTM), developed by Hochreiter and Schmidhuber in 1997, is an artificial neural net structure that unlike traditional RNNs doesn't have the vanishing gradient problem. It works even when there are long delays, and it can handle signals that have a mix of low and high frequency components. LSTM RNN outperformed other RNN and other sequence learning methods such as HMM in numerous applications such as language learning and connected handwriting recognition.

Bi-directional RNN

Invented by Schuster & Paliwal in 1997 bi-directional RNNs, or BRNNs, use a finite sequence to predict or label each element of the sequence based on both the past and the future context of the element. This is done by adding the outputs of two RNNs: one processing the sequence from left to right, the other one from right to left. The combined outputs are the predictions of the teacher-given target signals. This technique proved to be especially useful when combined with LSTM RNNs.

Hierarchical RNN

There are many instances of hierarchical RNN whose elements are connected in various ways to decompose hierarchical behavior into useful subprograms.

Stochastic Neural Networks

A stochastic neural network differs from a typical neural network because it introduces random variations into the network. In a probabilistic view of neural networks, such random variations can be viewed as a form of statistical sampling, such as Monte Carlo sampling.

Modular Neural Networks

Biological studies have shown that the human brain functions not as a single massive network, but as a collection of small networks. This realization gave birth to the concept of modular neural networks, in which several small networks cooperate or compete to solve problems.

Committee of Machines

A committee of machines (CoM) is a collection of different neural networks that together "vote" on a given example. This generally gives a much better result compared to other neural network models. Because neural networks suffer from local minima, starting with the same architecture and training but using different initial random weights often gives vastly different networks. A CoM tends to stabilize the result.

The CoM is similar to the general machine learning *bagging* method, except that the necessary variety of machines in the committee is obtained by training from different random starting weights rather than training on different randomly selected subsets of the training data.

Associative Neural Network (ASNN)

The ASNN is an extension of the *committee of machines* that goes beyond a simple/weighted average of different models. ASNN represents a combination of an ensemble of feedforward neural networks and the k-nearest neighbor technique (kNN). It uses the correlation between ensemble responses as a measure of distance amid the analyzed cases for the kNN. This corrects the bias of the neural network ensemble. An associative neural network has a memory that can coincide with the training set. If new data become available, the network instantly improves its predictive ability and provides data approximation (self-learn the data) without a need to retrain the ensemble. Another important feature of ASNN is the possibility to interpret neural network results by analysis of correlations between data cases in the space of models.

Physical Neural Network

A physical neural network includes electrically adjustable resistance material to simulate artificial synapses. Examples include the ADALINE neural network developed by Bernard Widrow in the 1960s and the memristor based neural network developed by Greg Snider of HP Labs in 2008 or the SyNAPSE Project by HRL Laboratories (HRL), Hewlett-Packard, and IBM Research.

Other Types of Networks

These special networks do not fit in any of the previous categories.

Holographic Associative Memory

Holographic associative memory represents a family of analog, correlation-based, associative, stimulus-response memories, where information is mapped onto the phase orientation of complex numbers operating.

Instantaneously Trained Networks

Instantaneously trained neural networks (ITNNs) were inspired by the phenomenon of short-term learning that seems to occur instantaneously. In these networks the weights of the hidden and the output layers are mapped directly from the training vector data. Ordinarily, they work on binary data, but versions for continuous data that require small additional processing are also available.

Spiking Neural Networks

Spiking neural networks (SNNs) are models which explicitly take into account the timing of inputs. The network input and output are usually represented as series of spikes (delta function or more complex shapes). SNNs have an advantage of being able to process information in the time domain (signals that vary over time). They are often implemented as recurrent networks. SNNs are also a form of pulse computer.

Spiking neural networks with axonal conduction delays exhibit polychronization, and hence could have a very large memory capacity.

Networks of spiking neurons—and the temporal correlations of neural assemblies in such networks—have been used to model figure/ground separation and region linking in the visual system.

In June 2005 IBM announced construction of a Blue Gene supercomputer dedicated to the simulation of a large recurrent spiking neural network.

Dynamic Neural Networks

Dynamic neural networks not only deal with nonlinear multivariate behaviour, but also include (learning of) time-dependent behaviour such as various transient phenomena and delay effects. Techniques to estimate a system process from observed data fall under the general category of system identification.

Cascading Neural Networks

Cascade correlation is an architecture and supervised learning algorithm developed by Scott Fahlman and Christian Lebiere. Instead of just adjusting the weights in a network of fixed topology, Cascade-Correlation begins with a minimal network, then automatically trains and adds new hidden units one by one, creating a multi-layer structure. Once a new hidden unit has been added to the network, its input-side weights are frozen. This unit then becomes a permanent feature-detector in the network, available for producing outputs or for creating other, more complex feature detectors. The Cascade-Correlation architecture has several advantages over existing algorithms: it learns very quickly, the network determines its own size and topology, it retains the structures it has built even if the training set changes, and it requires no back-propagation of error signals through the connections of the network.

Neuro-fuzzy Networks

A neuro-fuzzy network is a fuzzy inference system in the body of an artificial neural network. Depending on the *FIS* type, there are several layers that simulate the processes involved in a *fuzzy inference* like fuzzification, inference, aggregation and defuzzification. Embedding an *FIS* in a general structure of an *ANN* has the benefit of using available *ANN* training methods to find the parameters of a fuzzy system.

Compositional Pattern-producing Networks

Compositional pattern-producing networks (CPPNs) are a variation of ANNs which differ in their set of activation functions and how they are applied. While typical ANNs often contain only sigmoid functions (and sometimes Gaussian functions), CPPNs can include both types of functions and many others. Furthermore, unlike typical ANNs, CPPNs are applied across the entire space of possible inputs so that they can represent a complete image. Since they are compositions of functions, CPPNs in effect encode images at infinite resolution and can be sampled for a particular display at whatever resolution is optimal.

One-shot Associative Memory

This type of network can add new patterns without the need for re-training. It is done by creating a specific memory structure, which assigns each new pattern to an orthogonal plane using adjacently connected hierarchical arrays. The network offers real-time pattern recognition and high scalability; this however requires parallel processing and is thus best suited for platforms such as Wireless sensor networks (WSN), Grid computing, and GPGPUs.

Hierarchical Temporal Memory

Hierarchical temporal memory (HTM) is an online machine learning model developed by Jeff Hawkins and Dileep George of Numenta, Inc. that models some of the structural and algorithmic properties of the neocortex. HTM is a biomimetic model based on the memory-prediction theory of brain function described by Jeff Hawkins in his book *On Intelligence*. HTM is a method for discovering and inferring the high-level causes of observed input patterns and sequences, thus building an increasingly complex model of the world.

Jeff Hawkins states that HTM does not present any new idea or theory, but combines existing ideas to mimic the neocortex with a simple design that provides a large range of capabilities. HTM combines and extends approaches used in Bayesian networks, spatial and temporal clustering algorithms, while using a tree-shaped hierarchy of nodes that is common in neural networks.

Genetic Scale RNN

A RNN (often a LSTM) where a series is decomposed into a number of scales where every scale informs the primary length between two consecutive points. A first order scale consists of a normal RNN, a second order consists of all points separated by two indices and so on. The Nth order RNN connects the first and last node. The outputs from all the various Scales are treated as a CoM and the associated scores are used genetically for the next iteration.

Recurrent Neural Network

A recurrent neural network (RNN) is a class of artificial neural network where connections between units form a directed cycle. This creates an internal state of the network which allows it to exhibit dynamic temporal behavior. Unlike feedforward neural networks, RNNs can use their internal memory to process arbitrary sequences of inputs. This makes them applicable to tasks such as unsegmented connected handwriting recognition or speech recognition.

Architectures

Fully Recurrent Network

This is the basic architecture developed in the 1980s: a network of neuron-like units, each with a directed connection to every other unit. Each unit has a time-varying real-valued activation. Each connection has a modifiable real-valued weight. Some of the nodes are called input nodes, some output nodes, the rest hidden nodes. Most architectures below are special cases.

For supervised learning in discrete time settings, training sequences of real-valued input vectors become sequences of activations of the input nodes, one input vector at a time. At any given time step, each non-input unit computes its current activation as a nonlinear function of the weighted sum of the activations of all units from which it receives connections. There may be teacher-given target activations for some of the output units at certain time steps. For example, if the input sequence is a speech signal corresponding to a spoken digit, the final target output at the end of the sequence may be a label classifying the digit. For each sequence, its error is the sum of the deviations of all target signals from the corresponding activations computed by the network. For a training set of numerous sequences, the total error is the sum of the errors of all individual sequences.

In reinforcement learning settings, there is no teacher providing target signals for the RNN, instead a fitness function or reward function is occasionally used to evaluate the RNN's performance, which is influencing its input stream through output units connected to actuators affecting the environment.

Recursive Neural Networks

A recursive neural network is created by applying the same set of weights recursively over a differentiable graph-like structure, by traversing the structure in topological order. Such networks are typically also trained by the reverse mode of automatic differentiation. They were introduced to learn distributed representations of structure, such as logical terms. A special case of recursive neural networks is the RNN itself whose structure corresponds to a linear chain. Recursive neural networks have been applied to natural language processing. The Recursive Neural Tensor Network uses a tensor-based composition function for all nodes in the tree.

Hopfield Network

The Hopfield network is of historic interest although it is not a general RNN, as it is not designed to process sequences of patterns. Instead it requires stationary inputs. It is a RNN in which all connections are symmetric. Invented by John Hopfield in 1982, it guarantees that its dynamics will converge. If the connections are trained using Hebbian learning then the Hopfield network can perform as robust content-addressable memory, resistant to connection alteration.

A variation on the Hopfield network is the bidirectional associative memory (BAM). The BAM has two layers, either of which can be driven as an input, to recall an association and produce an output on the other layer.

Elman Networks and Jordan Networks

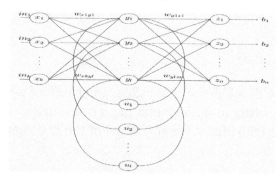

The Elman network

The following special case of the basic architecture above was employed by Jeff Elman. A three-layer network is used (arranged horizontally as x, y, and z in the illustration), with the addition of a set of "context units" (u in the illustration). There are connections from the middle (hidden) layer to these context units fixed with a weight of one. At each time step, the input is propagated in a standard feed-forward fashion, and then a learning rule is applied. The fixed back connections result in the context units always maintaining a copy of the previous values of the hidden units (since they propagate over the connections before the learning rule is applied). Thus the network can maintain a sort of state, allowing it to perform such tasks as sequence-prediction that are beyond the power of a standard multilayer perceptron.

Jordan networks, due to Michael I. Jordan, are similar to Elman networks. The context units are however fed from the output layer instead of the hidden layer. The context units in a Jordan network are also referred to as the state layer, and have a recurrent connection to themselves with no other nodes on this connection.

Elman and Jordan networks are also known as "simple recurrent networks" (SRN).

Elman network

$$h_t = \sigma_h(W_h x_t + U_h h_{t-1} + b_h)$$
$$y_t = \sigma_y(W_y h_t + b_y)$$

Jordan network

$$h_t = \sigma_h(W_h x_t + U_h y_{t-1} + b_h)$$
$$y_t = \sigma_y(W_y h_t + b_y)$$

Variables and functions

- x_t : input vector

- h_t : hidden layer vector

- y_t : output vector

- $W,\ U$ and b : parameter matrices and vector

- σ_h and σ_y : Activation functions

Echo State Network

The echo state network (ESN) is a recurrent neural network with a sparsely connected random hidden layer. The weights of output neurons are the only part of the network that can change and be trained. ESN are good at reproducing certain time series. A variant for spiking neurons is known as liquid state machines.

Neural History Compressor

The vanishing gradient problem of automatic differentiation or backpropagation in neural networks was partially overcome in 1992 by an early generative model called the neural history compressor, implemented as an unsupervised stack of recurrent neural networks (RNNs). The RNN at the input level learns to predict its next input from the previous input history. Only unpredictable inputs of some RNN in the hierarchy become inputs to the next higher level RNN which therefore recomputes its internal state only rarely. Each higher level RNN thus learns a compressed representation of the information in the RNN below. This is done such that the input sequence can be precisely reconstructed from the sequence representation at the highest level. The system effectively minimises the description length or the negative logarithm of the probability of the data. If there is a lot of learnable predictability in the incoming data sequence, then the highest level RNN can use supervised learning to easily classify even deep sequences with very long time intervals between important events. In 1993, such a system already solved a "Very Deep Learning" task that requires more than 1000 subsequent layers in an RNN unfolded in time.

It is also possible to distill the entire RNN hierarchy into only two RNNs called the "conscious" chunker (higher level) and the "subconscious" automatizer (lower level). Once the chunker has learned to predict and compress inputs that are still unpredictable by the automatizer, then the automatizer can be forced in the next learning phase to predict or imitate through special additional units the hidden units of the more slowly changing chunker. This makes it easy for the automatizer to learn appropriate, rarely changing memories across very long time intervals. This in turn helps the automatizer to make many of its once unpredictable inputs predictable, such that the chunker can focus on the remaining still unpredictable events, to compress the data even further.

Long Short-term Memory

Numerous researchers now use a deep learning RNN called the long short-term memory (LSTM) network, published by Hochreiter & Schmidhuber in 1997. It is a deep learning system that unlike traditional RNNs doesn't have the vanishing gradient problem. LSTM is normally augmented by recurrent gates called forget gates. LSTM RNNs prevent backpropagated errors from vanishing or exploding. Instead errors can flow backwards through unlimited numbers of virtual layers in LSTM RNNs unfolded in space. That is, LSTM can learn "Very Deep Learning" tasks that require memories of events that happened thousands or even millions of discrete time steps ago. Problem-specific LSTM-like topologies can be evolved. LSTM works even when there are long delays, and it can handle signals that have a mix of low and high frequency components.

Today, many applications use stacks of LSTM RNNs and train them by Connectionist Temporal Classification (CTC) to find an RNN weight matrix that maximizes the probability of the label sequences in a training set, given the corresponding input sequences. CTC achieves both alignment and recognition. Around 2007, LSTM started to revolutionize speech recognition, outperforming traditional models in certain speech applications. In 2009, CTC-trained LSTM was the first RNN to win pattern recognition contests, when it won several competitions in connected handwriting recognition. In 2014, the Chinese search giant Baidu used CTC-trained RNNs to break the Switchboard Hub5'00 speech recognition benchmark, without using any traditional speech processing methods. LSTM also improved large-vocabulary speech recognition, text-to-speech synthesis, also for Google Android, and photo-real talking heads. In 2015, Google's speech recognition reportedly experienced a dramatic performance jump of 49% through CTC-trained LSTM, which is now available through Google voice search to all smartphone users.

LSTM has also become very popular in the field of natural language processing. Unlike previous models based on HMMs and similar concepts, LSTM can learn to recognise context-sensitive languages. LSTM improved machine translation, Language Modeling and Multilingual Language

Processing. LSTM combined with convolutional neural networks (CNNs) also improved automatic image captioning and a plethora of other applications.

As of 2016, major technology companies are using LSTM networks as fundamental components in new products. Google uses LSTM for speech recognition on the smartphone, the smart assistant Allo, and Google Translate. Apple uses LSTM for the Quicktype function on the iPhone and for Siri. Amazon uses LSTM for Amazon Alexa.

Gated Recurrent Unit

Gated recurrent unit is one of the recurrent neural network introduced in 2014.

Bi-directional RNN

Invented by Schuster & Paliwal in 1997, bi-directional RNN or BRNN use a finite sequence to predict or label each element of the sequence based on both the past and the future context of the element. This is done by concatenating the outputs of two RNN, one processing the sequence from left to right, the other one from right to left. The combined outputs are the predictions of the teacher-given target signals. This technique proved to be especially useful when combined with LSTM RNN.

Continuous-time RNN

A continuous time recurrent neural network (CTRNN) is a dynamical systems model of biological neural networks. A CTRNN uses a system of ordinary differential equations to model the effects on a neuron of the incoming spike train.

For a neuron i in the network with action potential y_i the rate of change of activation is given by:

$$\tau_i \dot{y}_i = -y_i + \sum_{j=1}^{n} w_{ji} \sigma(y_j - \Theta_j) + I_i(t)$$

Where:

- τ_i : Time constant of postsynaptic node

- y_i : Activation of postsynaptic node

- $\dot{y}_i$: Rate of change of activation of postsynaptic node

- w_{ji} : Weight of connection from pre to postsynaptic node

- $\sigma(x)$: Sigmoid of x e.g. $\sigma(x) = 1/(1 + e^{-x})$.

- y_j : Activation of presynaptic node

- Θ_j : Bias of presynaptic node

- $I_i(t)$: Input (if any) to node

CTRNNs have frequently been applied in the field of evolutionary robotics, where they have been used to address, for example, vision, co-operation and minimally cognitive behaviour.

Note that by the Shannon sampling theorem, discrete time recurrent neural networks can be viewed as continuous time recurrent neural networks where the differential equation have transformed in an equivalent difference equation after that the postsynaptic node activation functions $y_i(t)$ have been low-pass filtered prior to sampling.

Hierarchical RNN

There are many instances of hierarchical RNN whose elements are connected in various ways to decompose hierarchical behavior into useful subprograms.

Recurrent Multilayer Perceptron

Generally, a Recurrent Multi-Layer Perceptron (RMLP) consists of a series of cascaded subnetworks, each of which consists of multiple layers of nodes. Each of these subnetworks is entirely feed-forward except for the last layer, which can have feedback connections among itself. Each of these subnets is connected only by feed forward connections.

Second Order RNN

Second order RNNs use higher order weights w_{ijk} instead of the standard w_{ij} weights, and inputs and states can be a product. This allows a direct mapping to a finite state machine both in training, stability, and representation. Long short-term memory is an example of this but has no such formal mappings or proof of stability.

Multiple Timescales Recurrent Neural Network (MTRNN) Model

MTRNN is a possible neural-based computational model that imitates to some extent the activity of the brain. It has the ability to simulate the functional hierarchy of the brain through self-organization that not only depends on spatial connection between neurons, but also on distinct types of neuron activities, each with distinct time properties. With such varied neuronal activities, continuous sequences of any set of behaviors are segmented into reusable primitives, which in turn are flexibly integrated into diverse sequential behaviors. The biological approval of such a type of hierarchy has been discussed on the memory-prediction theory of brain function by Jeff Hawkins in his book *On Intelligence*.

Pollack's Sequential Cascaded Networks

Neural Turing Machines

Neural Turing machine (NTMs) are a method of extending the capabilities of recurrent neural networks by coupling them to external memory resources, which they can interact with by attentional processes. The combined system is analogous to a Turing machine or Von Neumann architecture but is differentiable end-to-end, allowing it to be efficiently trained with gradient descent.

Neural Network Pushdown Automata

NNPDAs are similar to NTMs but tapes are replaced by analogue stacks that are differentiable and which are trained to control. In this way they are similar in complexity to recognizers of context free grammars (CFGs).

Bidirectional Associative Memory

First introduced by Kosko, BAM neural networks store associative data as a vector. The bi-directionality comes from passing information through a matrix and its transpose. Typically, bipolar encoding is preferred to binary encoding of the associative pairs. Recently, stochastic BAM models using Markov stepping were optimized for increased network stability and relevance to real-world applications.

Training

Gradient Descent

To minimize total error, gradient descent can be used to change each weight in proportion to the derivative of the error with respect to that weight, provided the non-linear activation functions are differentiable. Various methods for doing so were developed in the 1980s and early 1990s by Paul Werbos, Ronald J. Williams, Tony Robinson, Jürgen Schmidhuber, Sepp Hochreiter, Barak Pearlmutter, and others.

The standard method is called "backpropagation through time" or BPTT, and is a generalization of back-propagation for feed-forward networks, and like that method, is an instance of automatic differentiation in the reverse accumulation mode or Pontryagin's minimum principle. A more computationally expensive online variant is called "Real-Time Recurrent Learning" or RTRL, which is an instance of Automatic differentiation in the forward accumulation mode with stacked tangent vectors. Unlike BPTT this algorithm is *local in time but not local in space*.

In this context, *local in space* means that a unit's weight vector can be updated only using information stored in the connected units and the unit itself such that update complexity of a single unit is linear in the dimensionality of the weight vector. *Local in time* means that that the updates take place continually (on-line) and only depend on the most recent time step rather than on multiple time steps within a given time horizon as in BPTT. Biological neural networks appear to be local both with respect to time and space.

The downside of RTRL is that for recursively computing the partial derivatives, it has a time-complexity of O(number of hidden x number of weights) per time step for computing the Jacobian matrices, whereas BPTT only takes O(number of weights) per time step, at the cost, however, of storing all forward activations within the given time horizon.

There also is an online hybrid between BPTT and RTRL with intermediate complexity, and there are variants for continuous time. A major problem with gradient descent for standard RNN architectures is that error gradients vanish exponentially quickly with the size of the time lag between important events. The long short-term memory architecture together with a BPTT/RTRL hybrid learning method was introduced in an attempt to overcome these problems.

Moreover, the on-line algorithm called causal recursive BP (CRBP), implements and combines together BPTT and RTRL paradigms for locally recurrent network. It works with the most general locally recurrent networks. The CRBP algorithm can minimize the global error; this fact results in an improved stability of the algorithm, providing a unifying view on gradient calculation techniques for recurrent networks with local feedback.

An interesting approach to the computation of gradient information in RNNs with arbitrary architectures was proposed by Wan and Beaufays, is based on signal-flow graphs diagrammatic derivation to obtain the BPTT batch algorithm while, based on Lee theorem for networks sensitivity calculations, its fast online version was proposed by Campolucci, Uncini and Piazza.

Global Optimization Methods

Training the weights in a neural network can be modeled as a non-linear global optimization problem. A target function can be formed to evaluate the fitness or error of a particular weight vector as follows: First, the weights in the network are set according to the weight vector. Next, the network is evaluated against the training sequence. Typically, the sum-squared-difference between the predictions and the target values specified in the training sequence is used to represent the error of the current weight vector. Arbitrary global optimization techniques may then be used to minimize this target function.

The most common global optimization method for training RNNs is genetic algorithms, especially in unstructured networks.

Initially, the genetic algorithm is encoded with the neural network weights in a predefined manner where one gene in the chromosome represents one weight link, henceforth; the whole network is represented as a single chromosome. The fitness function is evaluated as follows: 1) each weight encoded in the chromosome is assigned to the respective weight link of the network; 2) the training set of examples is then presented to the network which propagates the input signals forward; 3) the mean-squared-error is returned to the fitness function; 4) this function will then drive the genetic selection process.

There are many chromosomes that make up the population; therefore, many different neural networks are evolved until a stopping criterion is satisfied. A common stopping scheme is: 1) when the neural network has learnt a certain percentage of the training data or 2) when the minimum value of the mean-squared-error is satisfied or 3) when the maximum number of training generations has been reached. The stopping criterion is evaluated by the fitness function as it gets the reciprocal of the mean-squared-error from each neural network during training. Therefore, the goal of the genetic algorithm is to maximize the fitness function, hence, reduce the mean-squared-error.

Other global (and/or evolutionary) optimization techniques may be used to seek a good set of weights such as simulated annealing or particle swarm optimization.

Related Fields and Models

RNNs may behave chaotically. In such cases, dynamical systems theory may be used for analysis.

Recurrent neural networks are in fact recursive neural networks with a particular structure: that of a linear chain. Whereas recursive neural networks operate on any hierarchical structure, combining child representations into parent representations, recurrent neural networks operate on the linear progression of time, combining the previous time step and a hidden representation into the representation for the current time step.

In particular, recurrent neural networks can appear as nonlinear versions of finite impulse response and infinite impulse response filters and also as a nonlinear autoregressive exogenous model (NARX).

Common RNN Libraries

- Apache Singa

- Caffe: Created by the Berkeley Vision and Learning Center (BVLC). It supports both CPU and GPU. Developed in C++, and has Python and MATLAB wrappers.

- Deeplearning4j: Deep learning in Java and Scala on multi-GPU-enabled Spark. A general-purpose deep learning library for the JVM production stack running on a C++ scientific computing engine. Allows the creation of custom layers. Integrates with Hadoop and Kafka.

- Keras

- Microsoft Cognitive Toolkit

- TensorFlow: Apache 2.0-licensed Theano-like library with support for CPU, GPU and Google's proprietary TPU, mobile

- Theano: The reference deep-learning library for Python with an API largely compatible with the popular NumPy library. Allows user to write symbolic mathematical expressions, then automatically generates their derivatives, saving the user from having to code gradients or backpropagation. These symbolic expressions are automatically compiled to CUDA code for a fast, on-the-GPU implementation.

- Torch: A scientific computing framework with wide support for machine learning algorithms, written in C and lua. The main author is Ronan Collobert, and it is now used at Facebook AI Research and Twitter.

Handwriting Recognition

Signature of country star, Tex Williams.

Handwriting recognition (or HWR) is the ability of a computer to receive and interpret intelligible handwritten input from sources such as paper documents, photographs, touch-screens and other devices. The image of the written text may be sensed "off line" from a piece of paper by optical scanning (optical character recognition) or intelligent word recognition. Alternatively, the movements of the pen tip may be sensed "on line", for example by a pen-based computer screen surface, a generally easier task as there are more clues available.

Handwriting recognition principally entails optical character recognition. However, a complete handwriting recognition system also handles formatting, performs correct segmentation into characters and finds the most plausible words.

Off-line Recognition

Off-line handwriting recognition involves the automatic conversion of text in an image into letter codes which are usable within computer and text-processing applications. The data obtained by this form is regarded as a static representation of handwriting. Off-line handwriting recognition is comparatively difficult, as different people have different handwriting styles. And, as of today, OCR engines are primarily focused on machine printed text and ICR for hand "printed" (written in capital letters) text.

Problem Domain Reduction Techniques

Narrowing the problem domain often helps increase the accuracy of handwriting recognition systems. A form field for a U.S. ZIP code, for example, would contain only the characters 0-9. This fact would reduce the number of possible identifications.

Primary techniques:

- Specifying specific character ranges

- Utilization of specialized forms

Character Extraction

Off-line character recognition often involves scanning a form or document written sometime in the past. This means the individual characters contained in the scanned image will need to be extracted. Tools exist that are capable of performing this step. However, there are several common imperfections in this step. The most common is when characters that are connected are returned as a single sub-image containing both characters. This causes a major problem in the recognition stage. Yet many algorithms are available that reduce the risk of connected characters.

Character Recognition

After the extraction of individual characters occurs, a recognition engine is used to identify the corresponding computer character. Several different recognition techniques are currently available.

Neural Networks

Neural network recognizers learn from an initial image training set. The trained network then makes the character identifications. Each neural network uniquely learns the properties that differentiate training images. It then looks for similar properties in the target image to be identified. Neural networks are quick to set up; however, they can be inaccurate if they learn properties that are not important in the target data.

Feature Extraction

Feature extraction works in a similar fashion to neural network recognizers. However, programmers must manually determine the properties they feel are important.

Some example properties might be:

- Aspect Ratio.
- Percent of pixels above horizontal half point
- Percent of pixels to right of vertical half point
- Number of strokes
- Average distance from image center
- Is reflected y axis
- Is reflected x axis

This approach gives the recognizer more control over the properties used in identification. Yet any system using this approach requires substantially more development time than a neural network because the properties are not learned automatically.

On-line Recognition and off-line Recognition

On-line handwriting recognition involves the automatic conversion of text as it is written on a special digitizer or PDA, where a sensor picks up the pen-tip movements as well as pen-up/pen-down switching. This kind of data is known as digital ink and can be regarded as a digital representation of handwriting. The obtained signal is converted into letter codes which are usable within computer and text-processing applications.

The elements of an on-line handwriting recognition interface typically include:

- a pen or stylus for the user to write with.
- a touch sensitive surface, which may be integrated with, or adjacent to, an output display.
- a software application which interprets the movements of the stylus across the writing surface, translating the resulting strokes into digital text. And an off-line recognition is the problem.

General Process

The process of online handwriting recognition can be broken down into a few general steps:

- preprocessing,
- feature extraction and
- classification

The purpose of preprocessing is to discard irrelevant information in the input data, that can negatively affect the recognition. This concerns speed and accuracy. Preprocessing usually consists of binarization, normalization, sampling, smoothing and denoising. The second step is feature extraction. Out of the two- or more-dimensional vector field received from the preprocessing algorithms, higher-dimensional data is extracted. The purpose of this step is to highlight important information for the recognition model. This data may include information like pen pressure, velocity or the changes of writing direction. The last big step is classification. In this step various models are used to map the extracted features to different classes and thus identifying the characters or words the features represent.

Hardware

Commercial products incorporating handwriting recognition as a replacement for keyboard input were introduced in the early 1980s. Examples include handwriting terminals such as the Pencept Penpad and the Inforite point-of-sale terminal. With the advent of the large consumer market for personal computers, several commercial products were introduced to replace the keyboard and mouse on a personal computer with a single pointing/handwriting system, such as those from Pen-Cept, CIC and others. The first commercially available tablet-type portable computer was the GRiD-Pad from GRiD Systems, released in September 1989. Its operating system was based on MS-DOS.

In the early 1990s, hardware makers including NCR, IBM and EO released tablet computers running the PenPoint operating system developed by GO Corp.. PenPoint used handwriting recognition and gestures throughout and provided the facilities to third-party software. IBM's tablet computer was the first to use the ThinkPad name and used IBM's handwriting recognition. This recognition system was later ported to Microsoft Windows for Pen Computing, and IBM's Pen for OS/2. None of these were commercially successful.

Advancements in electronics allowed the computing power necessary for handwriting recognition to fit into a smaller form factor than tablet computers, and handwriting recognition is often used as an input method for hand-held PDAs. The first PDA to provide written input was the Apple Newton, which exposed the public to the advantage of a streamlined user interface. However, the device was not a commercial success, owing to the unreliability of the software, which tried to learn a user's writing patterns. By the time of the release of the Newton OS 2.0, wherein the handwriting recognition was greatly improved, including unique features still not found in current recognition systems such as modeless error correction, the largely negative first impression had been made. After discontinuation of Apple Newton, the feature has been ported to Mac OS X 10.2 or later in form of Inkwell (Macintosh).

Palm later launched a successful series of PDAs based on the Graffiti recognition system. Graffiti improved usability by defining a set of "unistrokes", or one-stroke forms, for each character. This narrowed the possibility for erroneous input, although memorization of the stroke patterns did increase the learning curve for the user. The Graffiti handwriting recognition was found to infringe on a patent held by Xerox, and Palm replaced Graffiti with a licensed version of the CIC handwriting recognition which, while also supporting unistroke forms, pre-dated the Xerox patent. The court finding of infringement was reversed on appeal, and then reversed again on a later appeal. The parties involved subsequently negotiated a settlement concerning this and other patents Graffiti (Palm OS).

A Tablet PC is a special notebook computer that is outfitted with a digitizer tablet and a stylus, and allows a user to handwrite text on the unit's screen. The operating system recognizes the handwriting and converts it into typewritten text. Windows Vista and Windows 7 include personalization features that learn a user's writing patterns or vocabulary for English, Japanese, Chinese Traditional, Chinese Simplified and Korean. The features include a "personalization wizard" that prompts for samples of a user's handwriting and uses them to retrain the system for higher accuracy recognition. This system is distinct from the less advanced handwriting recognition system employed in its Windows Mobile OS for PDAs.

Although handwriting recognition is an input form that the public has become accustomed to, it has not achieved widespread use in either desktop computers or laptops. It is still generally accepted that keyboard input is both faster and more reliable. As of 2006, many PDAs offer handwriting input, sometimes even accepting natural cursive handwriting, but accuracy is still a problem, and some people still find even a simple on-screen keyboard more efficient.

Software

Initial software modules could understand print handwriting where the characters were separated. Author of the first applied pattern recognition program in 1962 was Shelia Guberman, then in Moscow. Commercial examples came from companies such as Communications Intelligence Corporation and IBM.

In the early 1990s, two companies, ParaGraph International, and Lexicus came up with systems that could understand cursive handwriting recognition. ParaGraph was based in Russia and founded by computer scientist Stepan Pachikov while Lexicus was founded by Ronjon Nag and Chris Kortge who were students at Stanford University. The ParaGraph CalliGrapher system was deployed in the Apple Newton systems, and Lexicus Longhand system was made available commercially for the PenPoint and Windows operating system. Lexicus was acquired by Motorola in 1993 and went on to develop Chinese handwriting recognition and predictive text systems for Motorola. ParaGraph was acquired in 1997 by SGI and its handwriting recognition team formed a P&I division, later acquired from SGI by Vadem. Microsoft has acquired CalliGrapher handwriting recognition and other digital ink technologies developed by P&I from Vadem in 1999.

Wolfram Mathematica (8.0 or later) also provides a handwriting or text recognition function Text Recognize.

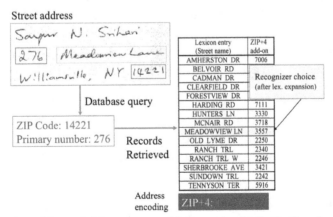

Method used for exploiting contextual information in the first handwritten address interpretation system developed by Sargur Srihari and Jonathan Hull

Handwriting Recognition has an active community of academics studying it. The biggest conferences for handwriting recognition are the International Conference on Frontiers in Handwriting Recognition (ICFHR), held in even-numbered years, and the International Conference on Document Analysis and Recognition (ICDAR), held in odd-numbered years. Both of these conferences are endorsed by the IEEE. Active areas of research include:

- Online Recognition

- Offline Recognition

- Signature Verification

- Postal-Address Interpretation

- Bank-Check Processing

- Writer Recognition

Results Since 2009

Since 2009, the recurrent neural networks and deep feedforward neural networks developed in the research group of Jürgen Schmidhuber at the Swiss AI Lab IDSIA have won several international handwriting competitions. In particular, the bi-directional and multi-dimensional Long short-term memory (LSTM) of Alex Graves et al. won three competitions in connected handwriting recognition at the 2009 International Conference on Document Analysis and Recognition (ICDAR), without any prior knowledge about the three different languages (French, Arabic, Persian) to be learned. Recent GPU-based deep learning methods for feedforward networks by Dan Ciresan and colleagues at IDSIA won the ICDAR 2011 offline Chinese handwriting recognition contest; their neural networks also were the first artificial pattern recognizers to achieve human-competitive performance on the famous MNIST handwritten digits problem of Yann LeCun and colleagues at NYU.

Speech Recognition

Speech recognition (SR) is the inter-disciplinary sub-field of computational linguistics that develops methodologies and technologies that enables the recognition and translation of spoken language into text by computers. It is also known as "automatic speech recognition" (ASR), "computer speech recognition", or just "speech to text" (STT). It incorporates knowledge and research in the linguistics, computer science, and electrical engineering fields.

Some SR systems use "training" (also called "enrollment") where an individual speaker reads text or isolated vocabulary into the system. The system analyzes the person's specific voice and uses it to fine-tune the recognition of that person's speech, resulting in increased accuracy. Systems that do not use training are called "speaker independent" systems. Systems that use training are called "speaker dependent".

Speech recognition applications include voice user interfaces such as voice dialing (e.g. "Call home"), call routing (e.g. "I would like to make a collect call"), domotic appliance control, search (e.g. find a podcast where particular words were spoken), simple data entry (e.g., entering a credit card number), preparation of structured documents (e.g. a radiology report), speech-to-text processing (e.g., word processors or emails), and aircraft (usually termed Direct Voice Input).

The term *voice recognition* or *speaker identification* refers to identifying the speaker, rather than what they are saying. Recognizing the speaker can simplify the task of translating speech in systems that have been trained on a specific person's voice or it can be used to authenticate or verify the identity of a speaker as part of a security process.

From the technology perspective, speech recognition has a long history with several waves of major innovations. Most recently, the field has benefited from advances in deep learning and big data. The advances are evidenced not only by the surge of academic papers published in the field, but more importantly by the worldwide industry adoption of a variety of deep learning methods in designing and deploying speech recognition systems. These speech industry players include Google, Microsoft, IBM, Baidu, Apple, Amazon, Nuance, SoundHound, IflyTek, CDAC many of which have publicized the core technology in their speech recognition systems as being based on deep learning.

History

In 1952 three Bell Labs researchers built a system for single-speaker digit recognition. Their system worked by locating the formants in the power spectrum of each utterance. The 1950s era technology was limited to single-speaker systems with vocabularies of around ten words.

Gunnar Fant developed the source-filter model of speech production and published it in 1960, which proved to be a useful model of speech production.

Unfortunately, funding at Bell Labs dried up for several years when, in 1969, the influential John Pierce wrote an open letter that was critical of speech recognition research. Pierce defunded speech recognition research at Bell Labs where no research on speech recognition was done until Pierce retired and James L. Flanagan took over.

Raj Reddy was the first person to take on continuous speech recognition as a graduate student at Stanford University in the late 1960s. Previous systems required the users to make a pause after each word. Reddy's system was designed to issue spoken commands for the game of chess.

Also around this time Soviet researchers invented the dynamic time warping (DTW) algorithm and used it to create a recognizer capable of operating on a 200-word vocabulary. The DTW algorithm processed the speech signal by dividing it into short frames, e.g. 10ms segments, and processing each frame as a single unit. Although DTW would be superseded by later algorithms, the technique of dividing the signal into frames would carry on. Achieving speaker independence was a major unsolved goal of researchers during this time period.

In 1971, DARPA funded five years of speech recognition research through its Speech Understanding Research program with ambitious end goals including a minimum vocabulary size of 1,000 words. BBN, IBM, Carnegie Mellon and Stanford Research Institute all participated in the program. The government funding revived speech recognition research that had been largely abandoned in the United States after John Pierce's letter.

Despite the fact that CMU's Harpy system met the original goals of the program, many predictions turned out to be nothing more than hype, disappointing DARPA administrators. This disappointment led to DARPA not continuing the funding. Several innovations happened during this time, such as the invention of beam search for use in CMU's Harpy system. The field also benefited from the discovery of several algorithms in other fields such as linear predictive coding and cepstral analysis.

During the late 1960s Leonard Baum developed the mathematics of Markov chains at the Institute for Defense Analysis. At CMU, Raj Reddy's students James Baker and Janet Baker began using the Hidden Markov Model (HMM) for speech recognition. James Baker had learned about HMMs from a summer job at the Institute of Defense Analysis during his undergraduate education. The use of HMMs allowed researchers to combine different sources of knowledge, such as acoustics, language, and syntax, in a unified probabilistic model.

Under Fred Jelinek's lead, IBM created a voice activated typewriter called Tangora, which could handle a 20,000 word vocabulary by the mid 1980s. Jelinek's statistical approach put less emphasis on emulating the way the human brain processes and understands speech in favor of using statistical modeling techniques like HMMs. (Jelinek's group independently discovered the application of HMMs to speech.) This was controversial with linguists since HMMs are too simplistic to account for many common features of human languages. However, the HMM proved to be a highly useful way for modeling speech and replaced dynamic time warping to become the dominant speech recognition algorithm in the 1980s. IBM had a few competitors including Dragon Systems founded by James and Janet Baker in 1982. The 1980s also saw the introduction of the n-gram language model. Katz introduced the back-off model in 1987, which allowed language models to use multiple length n-grams. During the same time, also CSELT was using HMM (the diphonies were studied since 1980) to recognize language like Italian. At the same time, CSELT led a series of European projects (Esprit I, II), and summarized the state-of-the-art in a book, later (2013) reprinted.

Much of the progress in the field is owed to the rapidly increasing capabilities of computers. At the end of the DARPA program in 1976, the best computer available to researchers was the PDP-10 with 4 MB ram. Using these computers it could take up to 100 minutes to decode just 30 seconds of speech. A few decades later, researchers had access to tens of thousands of times as much computing power. As the technology advanced and computers got faster, researchers began tackling harder problems such as larger vocabularies, speaker independence, noisy environments and conversational speech. In particular, this shifting to more difficult tasks has characterized DARPA funding of speech recognition since the 1980s. For example, progress was made on speaker independence first by training on a larger variety of speakers and then later by doing explicit speaker adaptation during decoding. Further reductions in word error rate came as researchers shifted acoustic models to be discriminative instead of using maximum likelihood models.

The 1990s saw the first introduction of commercially successful speech recognition technologies. By this point, the vocabulary of the typical commercial speech recognition system was larger than the average human vocabulary. Raj Reddy's former student, Xuedong Huang, developed the Sphinx-II system at CMU. The Sphinx-II system was the first to do speaker-independent, large vocabulary, continuous speech recognition and it had the best performance in DARPA's 1992 evaluation. Huang went on to found the speech recognition group at Microsoft in 1993. Raj Reddy's student Kai-Fu Lee joined Apple where, in 1992, he helped develop a speech interface prototype for the Apple computer known as Casper.

In 2000, Lernout & Hauspie acquired Dragon Systems and was an industry leader until an accounting scandal brought an end to the company in 2001. The L&H speech technology was bought by ScanSoft which became Nuance in 2005. Apple originally licensed software from Nuance to provide speech recognition capability to its digital assistant Siri.

21st Century

In the 2000s DARPA sponsored two speech recognition programs: Effective Affordable Reusable Speech-to-Text (EARS) in 2002 and Global Autonomous Language Exploitation (GALE). Four teams participated in the EARS program: IBM, a team led by BBN with LIMSI and Univ. of Pittsburgh, Cambridge University, and a team composed of ISCI, SRI and University of Washington. The GALE program focused on Arabic and Mandarin broadcast news speech. Google's first effort at speech recognition came in 2007 after hiring some researchers from Nuance. The first product was GOOG-411, a telephone based directory service. The recordings from GOOG-411 produced valuable data that helped Google improve their recognition systems. Google voice search is now supported in over 30 languages.

In the United States, the National Security Agency has made use of a type of speech recognition for keyword spotting since at least 2006. This technology allows analysts to search through large volumes of recorded conversations and isolate mentions of keywords. Recordings can be indexed and analysts can run queries over the database to find conversations of interest. Some government research programs focused on intelligence applications of speech recognition, e.g. DARPA's EARS's program and IARPA's Babel program.

In the early 2000s, speech recognition was still dominated by traditional approaches such as Hidden Markov Models combined with feedforward artificial neural networks. Today, however, many aspects of speech recognition have been taken over by a deep learning method called Long short-term memory (LSTM), a recurrent neural network published by Sepp Hochreiter & Jürgen Schmidhuber in 1997. LSTM RNNs avoid the vanishing gradient problem and can learn "Very Deep Learning" tasks that require memories of events that happened thousands of discrete time steps ago, which is important for speech. Around 2007, LSTM trained by Connectionist Temporal Classification (CTC) started to outperform traditional speech recognition in certain applications. In 2015, Google's speech recognition reportedly experienced a dramatic performance jump of 49% through CTC-trained LSTM, which is now available through Google Voice to all smartphone users.

The use of deep feedforward (non-recurrent) networks for acoustic modeling was introduced during later part of 2009 by Geoffrey Hinton and his students at University of Toronto and by Li Deng and colleagues at Microsoft Research, initially in the collaborative work between Microsoft and University of Toronto which was subsequently expanded to include IBM and Google (hence "The shared views of four research groups" subtitle in their 2012 review paper). A Microsoft research executive called this innovation "the most dramatic change in accuracy since 1979." In contrast to the steady incremental improvements of the past few decades, the application of deep learning decreased word error rate by 30%. This innovation was quickly adopted across the field. Researchers have begun to use deep learning techniques for language modeling as well.

In the long history of speech recognition, both shallow form and deep form (e.g. recurrent nets) of artificial neural networks had been explored for many years during 1980s, 1990s and a few years into the 2000s. But these methods never won over the non-uniform internal-handcrafting Gaussian mixture model/Hidden Markov model (GMM-HMM) technology based on generative models of speech trained discriminatively. A number of key difficulties had been methodologically analyzed in the 1990s, including gradient diminishing and weak temporal correlation structure in the neural predictive models. All these difficulties were in addition to the lack of big training data and big computing power in these early days. Most speech recognition researchers who understood such barriers hence subsequently moved away from neural nets to pursue generative modeling approaches until the recent resurgence of deep learning starting around 2009-2010 that had overcome all these difficulties. Hinton et al. and Deng et al. reviewed part of this recent history about how their collaboration with each other and then with colleagues across four groups (University of Toronto, Microsoft, Google, and IBM) ignited a renaissance of applications of deep feedforward neural networks to speech recognition.

Models, Methods, and Algorithms

Both acoustic modeling and language modeling are important parts of modern statistically-based speech recognition algorithms. Hidden Markov models (HMMs) are widely used in many systems. Language modeling is also used in many other natural language processing applications such as document classification or statistical machine translation.

Hidden Markov Models

Modern general-purpose speech recognition systems are based on Hidden Markov Models. These are statistical models that output a sequence of symbols or quantities. HMMs are used in speech recognition because a speech signal can be viewed as a piecewise stationary signal or a short-time stationary signal. In a short time-scale (e.g., 10 milliseconds), speech can be approximated as a stationary process. Speech can be thought of as a Markov model for many stochastic purposes.

Another reason why HMMs are popular is because they can be trained automatically and are simple and computationally feasible to use. In speech recognition, the hidden Markov model would output a sequence of n-dimensional real-valued vectors (with n being a small integer, such as 10), outputting one of these every 10 milliseconds. The vectors would consist of cepstral coefficients, which are obtained by taking a Fourier transform of a short time window of speech and decorrelating the spectrum using a cosine transform, then taking the first (most significant) coefficients. The hidden Markov model will tend to have in each state a statistical distribution that is a mixture of diagonal covariance Gaussians, which will give a likelihood for each observed vector. Each word, or (for more general speech recognition systems), each phoneme, will have a different output distribution; a hidden Markov model for a sequence of words or phonemes is made by concatenating the individual trained hidden Markov models for the separate words and phonemes.

Described above are the core elements of the most common, HMM-based approach to speech recognition. Modern speech recognition systems use various combinations of a number of standard techniques in order to improve results over the basic approach described above. A typical large-vocabulary system would need context dependency for the phonemes (so phonemes with different left and right context have different realizations as HMM states); it would use cepstral normalization to normalize for different speaker and recording conditions; for further speaker normalization it might use vocal tract length normalization (VTLN) for male-female normalization and maximum likelihood linear regression (MLLR) for more general speaker adaptation. The features would have so-called delta and delta-delta coefficients to capture speech dynamics and in addition might use heteroscedastic linear discriminant analysis (HLDA); or might skip the delta and delta-delta coefficients and use splicing and an LDA-based projection followed perhaps by heteroscedastic linear discriminant analysis or a global semi-tied co variance transform (also known as maximum likelihood linear transform, or MLLT). Many systems use so-called discriminative training techniques that dispense with a purely statistical approach to HMM parameter estimation and instead optimize some classification-related measure of the training data. Examples are maximum mutual information (MMI), minimum classification error (MCE) and minimum phone error (MPE).

Decoding of the speech (the term for what happens when the system is presented with a new utterance and must compute the most likely source sentence) would probably use the Viterbi algorithm to find the best path, and here there is a choice between dynamically creating a combination hidden Markov model, which includes both the acoustic and language model information, and combining it statically beforehand (the finite state transducer, or FST, approach).

A possible improvement to decoding is to keep a set of good candidates instead of just keeping the best candidate, and to use a better scoring function (re scoring) to rate these good candidates so that we may pick the best one according to this refined score. The set of candidates can be kept either as a list (the N-best list approach) or as a subset of the models (a lattice). Re scoring is usually done by trying to minimize the Bayes risk (or an approximation thereof): Instead of taking the source sentence with maximal probability, we try to take the sentence that minimizes the expectancy of a given loss function with regards to all possible transcriptions (i.e., we take the sentence that minimizes the average distance to other possible sentences weighted by their estimated probability). The loss function is usually the Levenshtein distance, though it can be different distances for specific tasks; the set of possible transcriptions is, of course, pruned to maintain tractability. Efficient algorithms have been devised to re score lattices represented as weighted finite state transducers with edit distances represented themselves as a finite state transducer verifying certain assumptions.

Dynamic Time Warping (DTW)-based Speech Recognition

Dynamic time warping is an approach that was historically used for speech recognition but has now largely been displaced by the more successful HMM-based approach.

Dynamic time warping is an algorithm for measuring similarity between two sequences that may vary in time or speed. For instance, similarities in walking patterns would be detected, even if in one video the person was walking slowly and if in another he or she were walking more quickly, or even if there were accelerations and deceleration during the course of one observation. DTW has been applied to video, audio, and graphics – indeed, any data that can be turned into a linear representation can be analyzed with DTW.

A well-known application has been automatic speech recognition, to cope with different speaking speeds. In general, it is a method that allows a computer to find an optimal match between two given sequences (e.g., time series) with certain restrictions. That is, the sequences are "warped" non-linearly to match each other. This sequence alignment method is often used in the context of hidden Markov models.

Neural Networks

Neural networks emerged as an attractive acoustic modeling approach in ASR in the late 1980s. Since then, neural networks have been used in many aspects of speech recognition such as phoneme classification, isolated word recognition, and speaker adaptation.

In contrast to HMMs, neural networks make no assumptions about feature statistical properties and have several qualities making them attractive recognition models for speech recognition. When used to estimate the probabilities of a speech feature segment, neural networks allow discriminative training in a natural and efficient manner. Few assumptions on the statistics of input features are made with neural networks. However, in spite of their effectiveness in classifying short-time units such as individual phones and isolated words, neural networks are rarely successful for continuous recognition tasks, largely because of their lack of ability to model temporal dependencies.

However, recently LSTM Recurrent Neural Networks (RNNs) and Time Delay Neural Networks(TDNN's) have been used which have been shown to be able to identify latent temporal dependencies and use this information to perform the task of speech recognition.

Deep Neural Networks and Denoising Autoencoders were also being experimented with to tackle this problem in an effective manner.

Due to the inability of feedforward Neural Networks to model temporal dependencies, an alternative approach is to use neural networks as a pre-processing e.g. feature transformation, dimensionality reduction, for the HMM based recognition.

Deep Feedforward and Recurrent Neural Networks

A deep feedforward neural network (DNN) is an artificial neural network with multiple hidden layers of units between the input and output layers. Similar to shallow neural networks, DNNs can model complex non-linear relationships. DNN architectures generate compositional models, where extra layers enable composition of features from lower layers, giving a huge learning capacity and thus the potential of modeling complex patterns of speech data.

A success of DNNs in large vocabulary speech recognition occurred in 2010 by industrial researchers, in collaboration with academic researchers, where large output layers of the DNN based on context dependent HMM states constructed by decision trees were adopted.

One fundamental principle of deep learning is to do away with hand-crafted feature engineering and to use raw features. This principle was first explored successfully in the architecture of deep autoencoder on the "raw" spectrogram or linear filter-bank features, showing its superiority over the Mel-Cepstral features which contain a few stages of fixed transformation from spectrograms. The true "raw" features of speech, waveforms, have more recently been shown to produce excellent larger-scale speech recognition results.

End-to-End Automatic Speech Recognition

Since 2014, there has been much research interest in end-to-end ASR. Traditional phonetic-based (i.e., all HMM-based model) approaches required separate components and training for the pronunciation, acoustic and language model. End-to-end models jointly learn all the components of the speech recognizer. This is valuable since it simplifies the training process and deployment process. For example, a n-gram language model is required for all HMM-based systems, and a typical n-gram language model often takes several gigabytes in memory making them impractical to deploy on mobile devices. Consequently, modern commercial ASR systems from Google and Apple (as of 2017) are deployed on the cloud and require a network connection as opposed to the device locally.

The first attempt of end-to-end ASR was with Connectionist Temporal Classification (CTC) based systems introduced by Alex Graves of Google DeepMind and Navdeep Jaitly of the University of Toronto in 2014. The model consisted of recurrent neural networks and a CTC layer. Jointly, the RNN-CTC model learns the pronunciation and acoustic model together, however it is incapable of learning the language due to conditional independence assumptions similar to a HMM. Consequently, CTC models can directly learn to map speech acoustics to English characters, but the models make many common spelling mistakes and must rely on a separate language model to clean up the transcripts. Later, Baidu expanded on the work with extremely large datasets and demonstrated some commercial success in Chinese Mandarin and English. In 2016, University of Oxford presented LipNet, the first end-to-end sentence-level lip reading model, using spatiotemporal convolutions coupled with an RNN-CTC architecture, surpassing human-level performance in a restricted grammar dataset.

An alternative approach to CTC-based models are attention-based models. Attention-based ASR models were introduced simultaneously by Chan et al. of Carnegie Mellon University and Google Brain and Bahdanaua et al. of the University of Montreal in 2016. The model named "Listen, Attend and Spell" (LAS), literally "listens" to the acoustic signal, pays "attention" to different parts of the signal and "spells" out the transcript one character at a time. Unlike CTC-based models, attention-based models do not have conditional-independence assumptions and can learn all the components of a speech recognizer including the pronunciation, acoustic and language model directly. This means, during deployment, there is no need to carry around a language model making it very practical for deployment onto applications with limited memory. By the end of 2016, the attention-based models have seen considerable success including outperforming the CTC models (with or without an external language model). Various extensions have been proposed since the original LAS model. Latent Sequence Decompositions (LSD) was proposed by Carnegie Mellon University, MIT and Google Brain to directly emit sub-word units which are more natural than English characters; University of Oxford and Google DeepMind extended LAS to "Watch, Listen, Attend and Spell" (WLAS) to handle lip reading surpassing human-level performance.

Applications

In-car Systems

Typically a manual control input, for example by means of a finger control on the steering-wheel, enables the speech recognition system and this is signalled to the driver by an audio prompt. Following the audio prompt, the system has a "listening window" during which it may accept a speech input for recognition.

Simple voice commands may be used to initiate phone calls, select radio stations or play music from a compatible smartphone, MP3 player or music-loaded flash drive. Voice recognition capabilities vary between car make and model. Some of the most recent car models offer natural-language speech recognition in place of a fixed set of commands, allowing the driver to use full sentences and common phrases. With such systems there is, therefore, no need for the user to memorize a set of fixed command words.

Health Care

Medical Documentation

In the health care sector, speech recognition can be implemented in front-end or back-end of the medical documentation process. Front-end speech recognition is where the provider dictates into a speech-recognition engine, the recognized words are displayed as they are spoken, and the dictator is responsible for editing and signing off on the document. Back-end or deferred speech recognition is where the provider dictates into a digital dictation system, the voice is routed through a speech-recognition machine and the recognized draft document is routed along with the original voice file to the editor, where the draft is edited and report finalized. Deferred speech recognition is widely used in the industry currently.

One of the major issues relating to the use of speech recognition in healthcare is that the American Recovery and Reinvestment Act of 2009 (ARRA) provides for substantial financial benefits to physicians who utilize an EMR according to "Meaningful Use" standards. These standards require that a substantial amount of data be maintained by the EMR (now more commonly referred to as an Electronic Health Record or EHR). The use of speech recognition is more naturally suited to the generation of narrative text, as part of a radiology/pathology interpretation, progress note or discharge summary: the ergonomic gains of using speech recognition to enter structured discrete data (e.g., numeric values or codes from a list or a controlled vocabulary) are relatively minimal for people who are sighted and who can operate a keyboard and mouse.

A more significant issue is that most EHRs have not been expressly tailored to take advantage of voice-recognition capabilities. A large part of the clinician's interaction with the EHR involves navigation through the user interface using menus, and tab/button clicks, and is heavily dependent on keyboard and mouse: voice-based navigation provides only modest ergonomic benefits. By contrast, many highly customized systems for radiology or pathology dictation implement voice "macros", where the use of certain phrases – e.g., "normal report", will automatically fill in a large number of default values and/or generate boilerplate, which will vary with the type of the exam – e.g., a chest X-ray vs. a gastrointestinal contrast series for a radiology system.

As an alternative to this navigation by hand, cascaded use of speech recognition and information extraction has been studied as a way to fill out a handover form for clinical proofing and sign-off. The results are encouraging, and the paper also opens data, together with the related performance benchmarks and some processing software, to the research and development community for studying clinical documentation and language-processing.

Therapeutic Use

Prolonged use of speech recognition software in conjunction with word processors has shown benefits to short-term-memory restrengthening in brain AVM patients who have been treated with resection. Further research needs to be conducted to determine cognitive benefits for individuals whose AVMs have been treated using radiologic techniques.

Military

High-performance Fighter Aircraft

Substantial efforts have been devoted in the last decade to the test and evaluation of speech recognition in fighter aircraft. Of particular note have been the US program in speech recognition for the Advanced Fighter Technology Integration (AFTI)/F-16 aircraft (F-16 VISTA), the program in France for Mirage aircraft, and other programs in the UK dealing with a variety of aircraft platforms. In these programs, speech recognizers have been operated successfully in fighter aircraft, with applications including: setting radio frequencies, commanding an auto-pilot system, setting steer-point coordinates and weapons release parameters, and controlling flight display.

Working with Swedish pilots flying in the JAS-39 Gripen cockpit, Englund (2004) found recognition deteriorated with increasing g-loads. The report also concluded that adaptation greatly improved the results in all cases and that the introduction of models for breathing was shown to improve recognition scores significantly. Contrary to what might have been expected, no effects of the broken English of the speakers were found. It was evident that spontaneous speech caused problems for the recognizer, as might have been expected. A restricted vocabulary, and above all, a proper syntax, could thus be expected to improve recognition accuracy substantially.

The Eurofighter Typhoon, currently in service with the UK RAF, employs a speaker-dependent system, requiring each pilot to create a template. The system is not used for any safety-critical or weapon-critical tasks, such as weapon release or lowering of the undercarriage, but is used for a wide range of other cockpit functions. Voice commands are confirmed by visual and/or aural feedback. The system is seen as a major design feature in the reduction of pilot workload, and even allows the pilot to assign targets to his aircraft with two simple voice commands or to any of his wingmen with only five commands.

Speaker-independent systems are also being developed and are under test for the F35 Lightning II (JSF) and the Alenia Aermacchi M-346 Master lead-in fighter trainer. These systems have produced word accuracy scores in excess of 98%.

Helicopters

The problems of achieving high recognition accuracy under stress and noise pertain strongly to the helicopter environment as well as to the jet fighter environment. The acoustic noise problem is actually more severe in the helicopter environment, not only because of the high noise levels but also because the helicopter pilot, in general, does not wear a facemask, which would reduce acoustic noise in the microphone. Substantial test and evaluation programs have been carried out in the past decade in speech recognition systems applications in helicopters, notably by the U.S. Army Avionics Research and Development Activity (AVRADA) and by the Royal Aerospace Establishment (RAE) in the UK. Work in France has included speech recognition in the Puma helicopter. There has also been much useful work in Canada. Results have been encouraging, and voice applications have included: control of communication radios, setting of navigation systems, and control of an automated target handover system.

As in fighter applications, the overriding issue for voice in helicopters is the impact on pilot effectiveness. Encouraging results are reported for the AVRADA tests, although these represent only a feasibility demonstration in a test environment. Much remains to be done both in speech recognition and in overall speech technology in order to consistently achieve performance improvements in operational settings.

Training Air Traffic Controllers

Training for air traffic controllers (ATC) represents an excellent application for speech recognition systems. Many ATC training systems currently require a person to act as a "pseudo-pilot", engaging in a voice dialog with the trainee controller, which simulates the dialog that the controller would have to conduct with pilots in a real ATC situation. Speech recognition and synthesis techniques offer the potential to eliminate the need for a person to act as pseudo-pilot, thus reducing training and support personnel. In theory, Air controller tasks are also characterized by highly structured speech as the primary output of the controller, hence reducing the difficulty of the speech recognition task should be possible. In practice, this is rarely the case. The FAA document 7110.65 details the phrases that should be used by air traffic controllers. While this document gives less than 150 examples of such phrases, the number of phrases supported by one of the simulation vendors speech recognition systems is in excess of 500,000.

The USAF, USMC, US Army, US Navy, and FAA as well as a number of international ATC training organizations such as the Royal Australian Air Force and Civil Aviation Authorities in Italy, Brazil, and Canada are currently using ATC simulators with speech recognition from a number of different vendors.

Telephony and Other Domain

ASR in the field of telephony is now commonplace and in the field of computer gaming and simulation is becoming more widespread. Despite the high level of integration with word processing in general personal computing. However, ASR in the field of document production has not seen the expected increases in use.

The improvement of mobile processor speeds made feasible the speech-enabled Symbian and Windows Mobile smartphones. Speech is used mostly as a part of a user interface, for creating predefined or custom speech commands. Leading software vendors in this field are: Google, Microsoft Corporation (Microsoft Voice Command), Digital Syphon (Sonic Extractor), LumenVox, Nuance Communications (Nuance Voice Control), Voci Technologies, VoiceBox Technology, Speech Technology Center, Vito Technologies (VITO Voice2Go), Speereo Software (Speereo Voice Translator), Verbyx VRX and SVOX.

Usage in Education and Daily Life

For language learning, speech recognition can be useful for learning a second language. It can teach proper pronunciation, in addition to helping a person develop fluency with their speaking skills.

Students who are blind or have very low vision can benefit from using the technology to convey words and then hear the computer recite them, as well as use a computer by commanding with their voice, instead of having to look at the screen and keyboard.

Students who are physically disabled or suffer from Repetitive strain injury/other injuries to the upper extremities can be relieved from having to worry about handwriting, typing, or working with scribe on school assignments by using speech-to-text programs. They can also utilize speech recognition technology to freely enjoy searching the Internet or using a computer at home without having to physically operate a mouse and keyboard.

Speech recognition can allow students with learning disabilities to become better writers. By saying the words aloud, they can increase the fluidity of their writing, and be alleviated of concerns regarding spelling, punctuation, and other mechanics of writing. Also.

Use of voice recognition software, in conjunction with a digital audio recorder and a personal computer running word-processing software has proven to be positive for restoring damaged short-term-memory capacity, in stroke and craniotomy individuals.

People with Disabilities

People with disabilities can benefit from speech recognition programs. For individuals that are Deaf or Hard of Hearing, speech recognition software is used to automatically generate a closed-captioning of conversations such as discussions in conference rooms, classroom lectures, and/or religious services.

Speech recognition is also very useful for people who have difficulty using their hands, ranging from mild repetitive stress injuries to involve disabilities that preclude using conventional computer input devices. In fact, people who used the keyboard a lot and developed RSI became an urgent early market for speech recognition. Speech recognition is used in deaf telephony, such as voicemail to text, relay services, and captioned telephone. Individuals with learning disabilities who have problems with thought-to-paper communication (essentially they think of an idea but it is processed incorrectly causing it to end up differently on paper) can possibly benefit from the software but the technology is not bug proof. Also the whole idea of speak to text can be hard for intellectually disabled person's due to the fact that it is rare that anyone tries to learn the technology to teach the person with the disability.

This type of technology can help those with dyslexia but other disabilities are still in question. The effectiveness of the product is the problem that is hindering it being effective. Although a kid may be able to say a word depending on how clear they say it the technology may think they are saying another word and input the wrong one. Giving them more work to fix, causing them to have to take more time with fixing the wrong word.

Performance

The performance of speech recognition systems is usually evaluated in terms of accuracy and speed. Accuracy is usually rated with word error rate (WER), whereas speed is measured with the real time factor. Other measures of accuracy include Single Word Error Rate (SWER) and Command Success Rate (CSR).

Speech recognition by machine is a very complex problem, however. Vocalizations vary in terms of accent, pronunciation, articulation, roughness, nasality, pitch, volume, and speed. Speech is distorted by a background noise and echoes, electrical characteristics. Accuracy of speech recognition may vary with the following:

- Vocabulary size and confusability

- Speaker dependence versus independence

- Isolated, discontinuous or continuous speech

- Task and language constraints

- Read versus spontaneous speech

- Adverse conditions

Accuracy

Accuracy of speech recognition may vary depending on the following factors:

- Error rates increase as the vocabulary size grows:

e.g. the 10 digits "zero" to "nine" can be recognized essentially perfectly, but vocabulary sizes of 200, 5000 or 100000 may have error rates of 3%, 7% or 45% respectively.

- Vocabulary is hard to recognize if it contains confusable words:

e.g. the 26 letters of the English alphabet are difficult to discriminate because they are confusable words (most notoriously, the E-set: "B, C, D, E, G, P, T, V, Z"); an 8% error rate is considered good for this vocabulary.

- Speaker dependence vs. independence:

A speaker-dependent system is intended for use by a single speaker.

A speaker-independent system is intended for use by any speaker (more difficult).

- Isolated, Discontinuous or continuous speech

With isolated speech, single words are used, therefore it becomes easier to recognize the speech.

With discontinuous speech full sentences separated by silence are used, therefore it becomes easier to recognize the speech as well as with isolated speech.

With continuous speech naturally spoken sentences are used, therefore it becomes harder to recognize the speech, different from both isolated and discontinuous speech.

- Task and language constraints

e.g. Querying application may dismiss the hypothesis "The apple is red."

e.g. Constraints may be semantic; rejecting "The apple is angry."

e.g. Syntactic; rejecting "Red is apple the."

Constraints are often represented by a grammar.

- Read vs. Spontaneous Speech

When a person reads it's usually in a context that has been previously prepared, but when a person uses spontaneous speech, it is difficult to recognize the speech because of the disfluencies (like "uh" and "um", false starts, incomplete sentences, stuttering, coughing, and laughter) and limited vocabulary.

- Adverse conditions

Environmental noise (e.g. Noise in a car or a factory)

Acoustical distortions (e.g. echoes, room acoustics)

Speech recognition is a multi-levelled pattern recognition task.

- Acoustical signals are structured into a hierarchy of units;

e.g. Phonemes, Words, Phrases, and Sentences;

- Each level provides additional constraints;

e.g. Known word pronunciations or legal word sequences, which can compensate for errors or uncertainties at lower level;

- This hierarchy of constraints are exploited;

By combining decisions probabilistically at all lower levels, and making more deterministic decisions only at the highest level, speech recognition by a machine is a process broken into several phases. Computationally, it is a problem in which a sound pattern has to be recognized or classified into a category that represents a meaning to a human. Every acoustic signal can be broken in smaller more basic sub-signals. As the more complex sound signal is broken into the smaller sub-sounds, different levels are created, where at the top level we have complex sounds, which are made of simpler sounds on lower level, and going to lower levels even more, we create more basic and shorter and simpler sounds. The lowest level, where the sounds are the most fundamental, a machine would check for simple and more probabilistic rules of what sound should represent. Once these sounds are put together into more complex sound on upper level, a new set of more deterministic rules should predict what new complex sound should represent. The most upper level of a deterministic rule should figure out the meaning of complex expressions. In order to expand our knowledge about speech recognition we need to take into a consideration neural networks. There are four steps of neural network approaches:

- Digitize the speech that we want to recognize

For telephone speech the sampling rate is 8000 samples per second;

- Compute features of spectral-domain of the speech (with Fourier transform);

computed every 10 ms, with one 10 ms section called a frame;

Analysis of four-step neural network approaches can be explained. Sound is produced by air (or some other medium) vibration, which we register by ears, but machines by receivers. Basic sound creates a wave which has 2 descriptions; Amplitude (how strong is it), and frequency (how often it vibrates per second).

The sound waves can be digitized: Sample a strength at short intervals like in picture above to get bunch of numbers that approximate at each time step the strength of a wave. Collection of these numbers represent analog wave. This new wave is digital. Sound waves are complicated because they superimpose one on top of each other. Like the waves would. This way they create odd-looking waves. For example, if there are two waves that interact with each other we can add them which creates new odd-looking wave.

- Neural network classifies features into phonetic-based categories;

Given basic sound blocks that a machine digitized, one has a bunch of numbers which describe a wave and waves describe words. Each frame has a unit block of sound, which are broken into basic sound waves and represented by numbers which, after Fourier Transform, can be statistically evaluated to set to which class of sounds it belongs. The nodes in the figure on a slide represent a feature of a sound in which a feature of a wave from the first layer of nodes to the second layer of nodes based on statistical analysis. This analysis depends on programmer's instructions. At this point, a second layer of nodes represents higher level features of a sound input which is again statistically evaluated to see what class they belong to. Last level of nodes should be output nodes that tell us with high probability what original sound really was.

- Search to match the neural-network output scores for the best word, to determine the word that was most likely uttered.

References

- J. Weng, N. Ahuja and T. S. Huang, "Learning recognition and segmentation using the Cresceptron," International Journal of Computer Vision, vol. 25, no. 2, pp. 105-139, Nov. 1997

- David H. Hubel and Torsten N. Wiesel (2005). Brain and visual perception: the story of a 25-year collaboration. Oxford University Press US. p. 106. ISBN 978-0-19-517618-6

- Kleene, S.C. (1956). "Representation of Events in Nerve Nets and Finite Automata". Annals of Mathematics Studies (34). Princeton University Press. pp. 3–41. Retrieved 2017-06-17

- Dreyfus, Stuart E. (1990-09-01). "Artificial neural networks, back propagation, and the Kelley-Bryson gradient procedure". Journal of Guidance, Control, and Dynamics. 13 (5): 926–928. ISSN 0731-5090. doi:10.2514/3.25422

- Rumelhart, D.E; McClelland, James (1986). Parallel Distributed Processing: Explorations in the Microstructure of Cognition. Cambridge: MIT Press. ISBN 978-0-262-63110-5

- Schmidhuber, J. (2015). "Deep Learning in Neural Networks: An Overview". Neural Networks. 61: 85–117. arXiv:1404.7828. doi:10.1016/j.neunet.2014.09.003

- Venkatachalam, V; Selvan, S. (2007). "Intrusion Detection using an Improved Competitive Learning Lamstar Network". International Journal of Computer Science and Network Security. 7 (2): 255–263

- Griewank, Andreas; Walther, Andrea (2008). Evaluating Derivatives: Principles and Techniques of Algorithmic Differentiation, Second Edition. SIAM. ISBN 978-0-89871-776-1

- Fan, Y.; Qian, Y.; Xie, F.; Soong, F. K. (2014). "TTS synthesis with bidirectional LSTM based Recurrent Neural Networks". ResearchGate. Retrieved 2017-06-13

- Mitchell, T; Beauchamp, J (1988). "Bayesian Variable Selection in Linear Regression". Journal of the American Statistical Association. 83 (404): 1023–1032. doi:10.1080/01621459.1988.10478694

- Rumelhart, David E.; Hinton, Geoffrey E.; Williams, Ronald J. "Learning representations by back-propagating errors". Nature. 323 (6088): 533–536. doi:10.1038/323533a0

- Graupe, Daniel (7 July 2016). Deep Learning Neural Networks: Design and Case Studies. World Scientific Publishing Co Inc. pp. 57–110. ISBN 978-981-314-647-1

- Adrian, Edward D. (1926). "The impulses produced by sensory nerve endings". The Journal of Physiology. 61 (1): 49–72. doi:10.1113/jphysiol.1926.sp002273

- Dewdney, A. K. (1 April 1997). Yes, we have no neutrons: an eye-opening tour through the twists and turns of bad science. Wiley. p. 82. ISBN 978-0-471-10806-1

- Burgess, Matt. "DeepMind's AI learned to ride the London Underground using human-like reason and memory". WIRED UK. Retrieved 2016-10-19

- Gers, Felix A.; Schmidhuber, Jürgen (2001). "LSTM Recurrent Networks Learn Simple Context Free and Context Sensitive Languages". IEEE TNN. 12 (6): 1333–1340. doi:10.1109/72.963769

Permissions

Index